An Introduction to Particle Accelerators

An Introduction to Particle Accelerators

E. J. N. WILSON
Head of the CERN Accelerator School
Geneva

OXFORD
UNIVERSITY PRESS

Great Clarendon Street, Oxford OX2 6DP

Oxford University Press is a department of the University of Oxford. It furthers the University's objective of excellence in research, scholarship, and education by publishing worldwide in

Oxford New York

Athens Auckland Bangkok Bogotá Buenos Aires Cape Town Chennai Dar es Salaam Delhi Florence Hong Kong Istanbul Karachi Kolkata Kuala Lumpur Madrid Melbourne Mexico City Mumbai Nairobi Paris São Paulo Shanghai Singapore Taipei Tokyo Toronto Warsaw

with associated companies in Berlin Ibadan

Published in the United States
by Oxford University Press Inc., New York

First published 2001

A catalogue record of this book is available from the British Library

Library of Congress Cataloging in Publication Data

Wilson, E.J.N. (Edward J.N.)
An introduction to particle accelerators/ E.J.N. Wilson
Includes bibliographical references and index
1. Particle accelerators. I. Title
QC787.P3 W57 2001 539.7′3–dc21 2001018540

ISBN 0 19 850829 8 (Pbk.)
ISBN 0 19 852054 9 (Hbk.)

Typeset by
Newgen Imaging Systems (P) Ltd., Chennai, India
Printed in Great Britain
on acid-free paper by
T.J. International Ltd.,
Padstow, Cornwall

To my dear wife—Monika

Preface

Since their invention in the 1930s, particle accelerators have grown in size to become the largest research tools at the disposal of the scientist—some even visible from spacecraft. At particle physics laboratories: CERN in Geneva, FNAL in Illinois, and DESY in Hamburg, teams of several hundred physicists and engineers construct huge rings of electromagnets and microwave accelerating cavities—each a densely packed mosaic of sophisticated components. There is hardly a major engineering project, with the possible exception of space research, which combines as many of today's ongoing technologies.

Apart from a few of these large accelerators, whose principal purpose is to further man's quest to understand his origin and that of the universe, there is a second and more numerous wave of smaller accelerators built for more practical purposes. These are used in industrial processes, in medicine for diagnosis and therapy, and to research into the physical and molecular structure of materials.

This book is intended to provide students with an understanding of the physics of accelerators—large and small—and to convey the flavour of their technology and applications. The first stumbling block for students of a new field is often the lack of a simple introduction which reveals the physical principles and which explains the concept and language of the subject. The book sets out to remedy this. It does not pretend to be an exhaustive reference but follows a pattern of learning which best matches the needs of a graduate engineer or physicist confronting the subject for the first time.

The subject matter falls roughly into three parts. After a chapter on history we find out how to design the patterns of bending and focusing magnets of a synchrotron, to maintain a stable circulating beam and how to accelerate these particles to a high energy. We then learn how to reach the highest possible intensity in the face of the many instabilities which may afflict the beam and which provide a formidable mathematical challenge. The last chapters are devoted to present-day colliders, applications of accelerators and a discussion of prospects for inventing new kinds of accelerators. There are many examples to help the student through the more theoretical chapters.

Acknowledgements

The author is deeply indebted to a number of colleagues who helped in the preparation of this manuscript—particularly Chris Prior, who read all the chapters and made many helpful suggestions and corrections, to Simon Baird, Oscar Barbalat, Helmut Burkhardt, Albert Hofmann, Kjell Johnson, Michel Martini,

Dieter Möhl, Werner Pirkl, Luigi Rinolfi and Wolfgang Schnell who read various chapters, and to Barbara Strasser, for her skill and patience in preparing the manuscript. Many of the lecturers at the CERN Accelerator School contributed by providing a large body of authoritative material used as a basis for this book. Finally I would like to thank my son Alex, whose encouragement was the key to completion.

Geneva E.J.N.W.
September 2000

Contents

1 History of accelerators

1.1 Overview of the history

The years around 1930 were exciting times for the inventors of accelerators. It was suddenly realized that the key to sustained acceleration was to use an electromagnetic field which varied in time. Particles might be accelerated indefinitely if they circulated in a rising magnetic field or if they passed many times through a relatively weak alternating potential difference between two electrodes. Three basic accelerator types, the betatron, the linac, and the cyclotron were invented opening up the possibility of almost indefinite acceleration. This led to the construction of a series of magnetic rings of larger and larger diameter to accelerate particles to energies which increased by an order of magnitude per decade.

We trace the progress of the field in Fig. 1.1, which shows how electron and proton accelerators developed from modest beginnings to become the most powerful tools available today for the study of physics. The motivation to strive for higher energies came from quantum mechanics, which describes particles as waves whose length is related to the momentum of the particle by De Broglie's expression:

$$\lambda = \frac{h}{p}.$$

Higher momentum brings shorter wavelengths and the capability to reveal ever finer detail in the structure of fundamental particles. Just as an electron microscope has better resolution than its optical counterpart, so the particle accelerator takes the quest to understand the finest details of sub-nuclear matter a stage further. Hand-in-hand with the understanding of smaller and smaller structures came the discovery of a whole series of ever more massive particles requiring, according to Einstein's $E = mc^2$, more and more energetic particles to produce them.

As particles are accelerated to energies many times their rest mass, the classical relations between velocity, momentum, and energy have to be abandoned in favour of the definitions of special relativity. In this description, although velocity saturates asymptotically—always just below the velocity of light—momentum and energy continue to increase as particles are accelerated. However, the radius of the circular orbit which the particle can follow in a magnetic field also increases. Each new accelerator has tended to be an order of magnitude larger

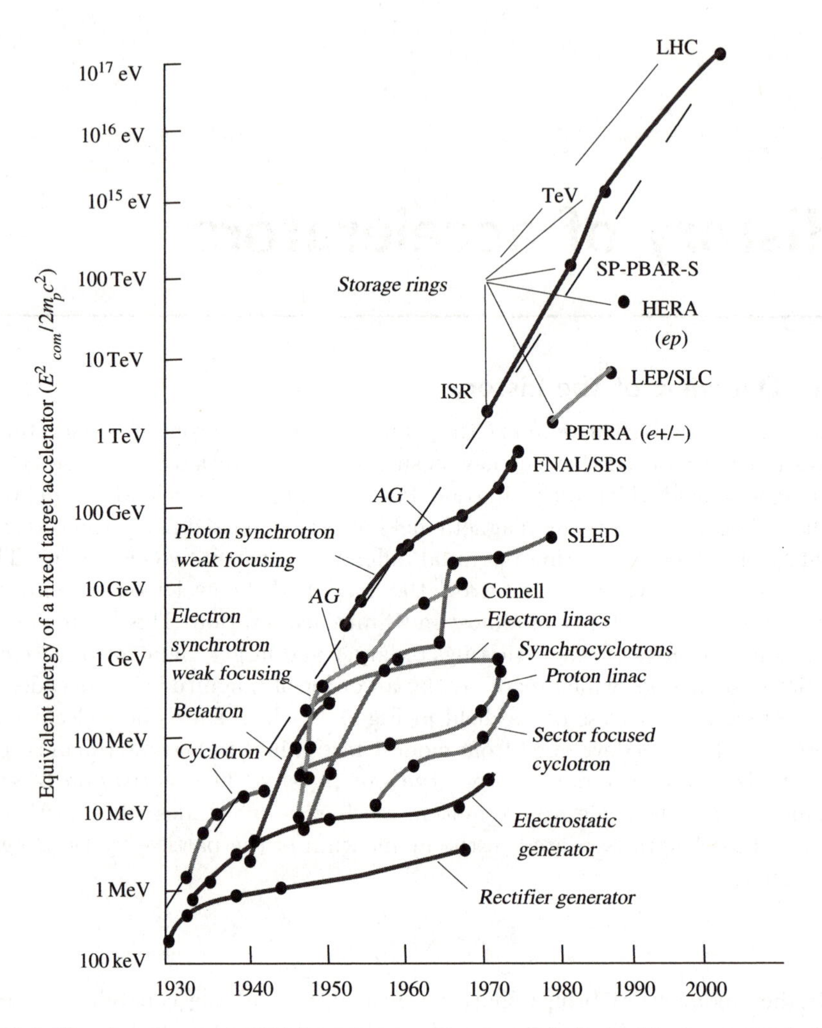

Fig. 1.1 The development of high-energy accelerators labelled with the technological advances which contributed to this progress.

in energy and radius than its predecessor, a machine which itself had often been regarded by its constructors as the ultimate in size.

Early physics experiments used beams of accelerated particles to bombard fixed targets. In such an encounter only a fraction of the energy is available to create new and interesting particles; the rest, in the form of the kinetic energy of the emerging particles, simply ensures that momentum is conserved. On the other hand, if two beams of particles can be arranged to collide head-on, there is no momentum to be conserved and all the energy of the incoming projectiles is available for particle creation. Although this was understood from early days, it was only latterly that two beams could be made dense enough to give a useful collision rate. But as beam instabilities were mastered, fixed target accelerators gave way to storage rings in which two counter-rotating beams continuously meet head-on.

The guide field of accelerators was limited to the 2 T maximum field of room-temperature magnets and accelerators grew to be several kilometres in circumference. In more recent years superconducting magnets have allowed us to increase the field to 4 T, and more recently 8 T, which gives a temporary respite from the expansionist tendency. Let us now return to the earliest days of accelerators.

1.2 Electrostatic accelerators

The reader is probably already familiar with the principle of the electrostatic accelerator which forms the electron gun of an ordinary TV set (Fig. 1.2). Electrons flow from a heated filament at earth potential at the cathode towards a positive anode plate and shoot through a small hole towards the screen. They acquire an energy (in joules) which is just their charge multiplied by the potential difference between cathode and anode. In the accelerator world this potential difference is used as the measure of energy. If a kilovolt is applied to the gun, the electron's energy is simply one thousand electron volts, 1 keV. To express this in joules one simply multiplies by the electron's charge.

In the years immediately after the end of the First World War, there were no accelerators. The projectiles used in Rutherford's pioneering scattering experiments were alpha particles from radioactive decay. No doubt he would dearly have liked to have had an accelerator of a few million electron volts, MeV, as a controlled source, but although the nineteenth century had produced a number of electrostatic high-voltage generators, they were unpredictable in performance and electrical breakdown became a serious problem above a few tens of kV.

Among the first high-voltage generators to approach 1 MeV was one built by Cockcroft and Walton (1932, 1934) to accelerate particles for their fission experiments. Their staircase of diode rectifiers is still used today to apply a

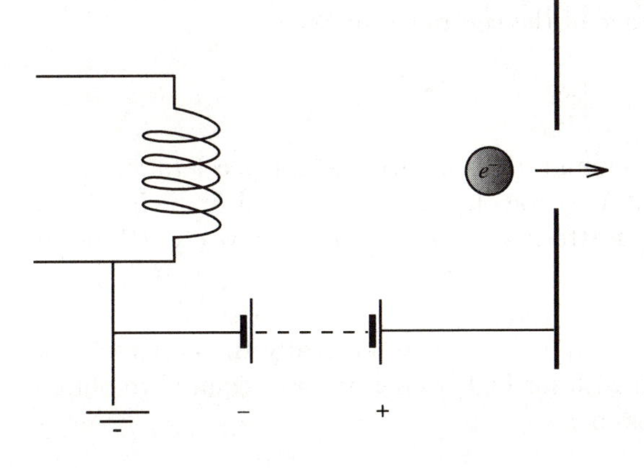

Fig. 1.2 An electron gun in a cathode ray tube.

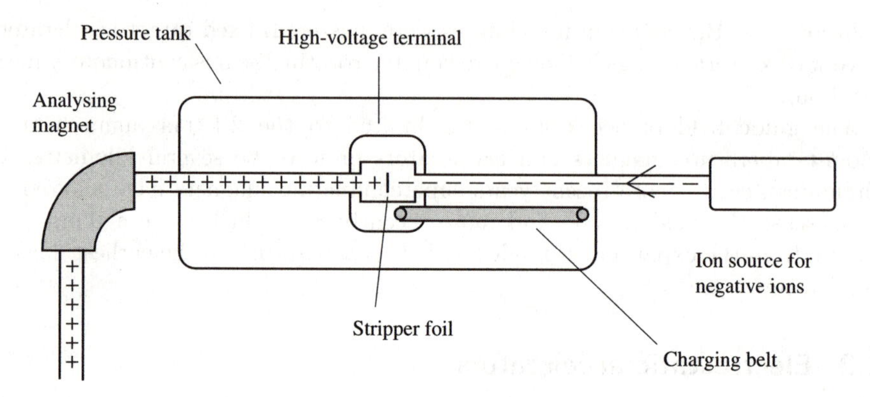

Fig. 1.3 Tandem van de Graaff accelerator.

high voltage to the ion or proton source at the beginning of many linacs and synchrotrons.

The early 1930s, also saw the invention by R. J. van de Graaff (1931), when still a Rhodes scholar at Oxford, of an electrostatic generator which used a moving belt to carry charge to the high-voltage terminal until it reaches a potential of several MV—rather as the sectors on the disc of a Wimshurst machine charge up the spheres of the spark gap. van de Graaff accelerators have proved a useful source of low-energy particles to this day but are inevitably limited by problems of voltage breakdown.

It is possible in theory to chain together several electrostatic accelerators, each with its cathode connected to the anode of the next, but each stage increases the potential between the ends of the device and between the ends and ground.

The nearest we have come to building such a device is the 'tandem' van de Graaff. Figure 1.3 shows two back-to-back machines sharing a common central high-voltage terminal but with their entrance and exit ports at ground potential. Positively charged ions become negative for the second half of their journey as they are stripped of electrons by a foil inside the central terminal. The negative ions are accelerated in the reversed field of the second stage from positive to ground and reach twice the voltage of the central terminal.

1.3 The ray transformer

The first attempt to overcome the limitations of electrostatic acceleration came from the inventive mind of Rolf Wideröe. In 1919, while still at high school, Wideröe read of Rutherford's scattering studies and later wrote (Wideröe 1994).

> 'It was clear to me that natural alpha rays were not really the best tools for the task; many more particles with far higher energy were required to obtain a greater number of nuclear fissions.'

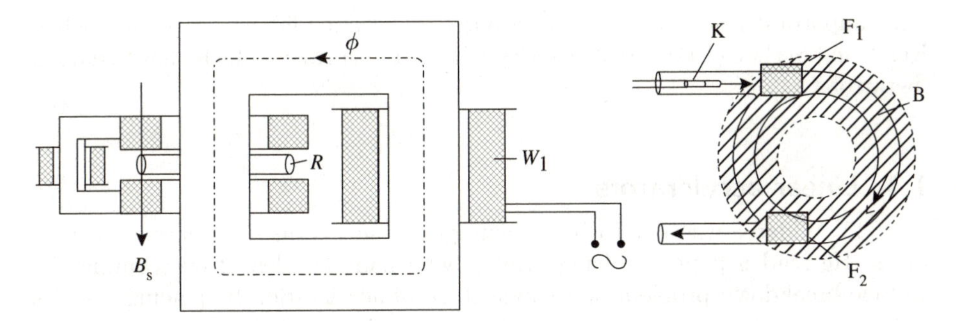

Fig. 1.4 Wideröe's sketch of the ray transformer (from Wideröe 1928).

In 1923, as he started his studies at Karlsruhe Technical University, he began wondering if electrons in an evacuated ring would flow in the same way as the electrons in copper in the secondary winding of a transformer (Wideröe 1923–1928) and hit upon the novel idea of the first circular accelerator. His notebooks of that time contain sketches of this device which he called a 'ray transformer', the precursor of the 'betatron'.

These sketches show a toroidal beam tube, R, placed in the gap between the parallel poles or faces of a small electromagnet (on the left in Fig. 1.4). The magnet is in the form of a 'C' and the field between the poles, B_s, guides particles in a circular orbit in the mid-plane between the poles. A circular hole is cut in each pole through which the yoke of the transformer passes linking the beam tube. The primary winding of the transformer, labelled W_1, is excited with alternating voltage from the mains. The beam tube is placed where one would normally expect the transformer's secondary winding to be and the beam of electrons within it carries the induced current. The windings of the C-magnet and of the primary of the transformer W_1 give independent control of the guide field and accelerating flux.

The ray transformer, unlike almost all accelerators that followed, relied entirely upon the inductive effect of a varying magnetic field and it is the rate of change of flux, ϕ, in the yoke which induces an accelerating voltage around the beam's path.

Wideröe used Einstein's newly discovered theory of special relativity to describe correctly the motion of particles close to the speed of light. He calculated that electrons circulating in a ring of only 10 or 20 cm diameter could reach several MeV within one quarter wave of the AC excitation of the transformer. He also found an important principle which ensures that the beam radius does not change as it accelerates. The total flux linking the beam must be exactly twice that enclosed by the beam circulating in a uniform dipole field. The extra flux is carried by the transformer's yoke which links the beam.

Unfortunately, Wideröe was dissuaded from building the ray transformer by difficulties with surface fields and by his professor who wrongly assumed the beam would be lost because of gas scattering. However, his ray transformer and the 2 : 1 ratio of accelerating to guide flux, now known as the Wideröe principle, were important discoveries and were put into practice fifteen years later when Kerst and Serber (1941) built a series of ray transformers which they renamed 'betatrons'.

1.4 Linear accelerators

In 1927, Wideröe was not to be discouraged from his quest to accelerate particles. He read a paper by Ising (1924), who had the idea of overcoming the voltage breakdown problem of a single stage of acceleration by placing a series of hollow cylindrical electrodes one after another in a straight line to form what today we would call a 'drift tube linac' or linear accelerator. In Ising's sketch a pulsed waveform is applied to each drift tube in turn to set up an accelerating field in each gap. The particles are shielded inside the drift tubes while the pulse is applied. Wideröe's contribution was to realize that an oscillating potential applied to one drift tube flanked by two others which are earthed, would accelerate at both gaps provided the oscillator's phase changes by 180° during the flight time between gaps.

He built a three-tube model which accelerated sodium ions (Wideröe 1928) and this was accepted for his thesis. However, although he realized that one might extend such a series of tubes indefinitely he did not take the idea any further as he was due to start his professional employment designing high-voltage circuit breakers. Between 1931 and 1934, D. Sloan and E. O. Lawrence at Berkeley took up Wideröe's idea and constructed mercury ion linacs with as many as 30 drift tubes but these were not used for nuclear research.

It was much later, in the mid-1940s, that L. W. Alvarez (1946) at the Radiation Laboratory of the University of California started to build the first serious proton linac. By this time suitable high-power, high-frequency oscillators had become available to meet the needs of war-time radar development. Figure 1.5 shows an Alvarez linac—a copper-lined cylinder excited by a radio transmitter. As in Wideröe's linac, particles gain energy from the accelerating potential differences between the ends of the drift tube, but the phase shift between drift tube gaps is 360°. Alternate tubes need not be earthed and each gap appears to the particle to be an identical field gradient which accelerates particles from left to right. The particles are protected from the decelerating phase while inside the metallic drift tubes.

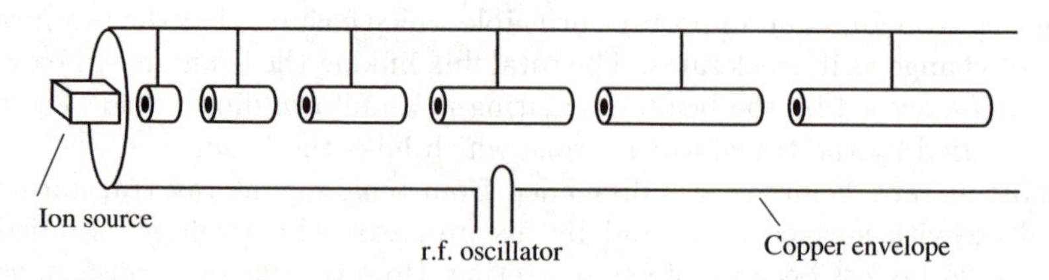

Fig. 1.5 The concept of the Alvarez linac (from Livingood 1961).

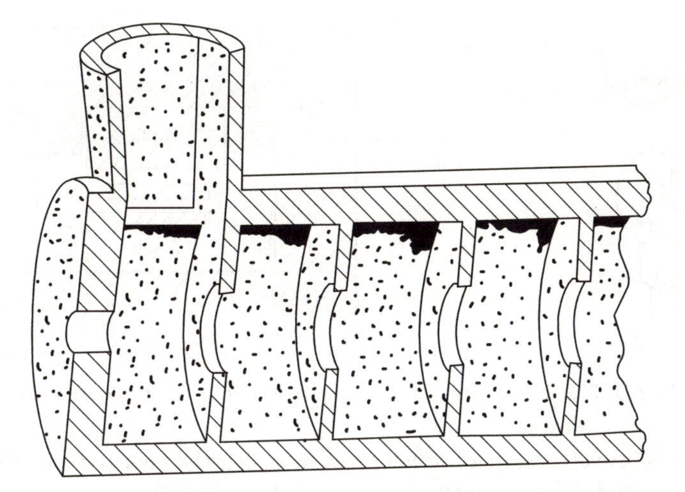

Fig. 1.6 Iris-loaded structure (from Lapostolle 1986). The 'chimney' is the input waveguide.

As one might expect, the distance between gaps increases as the particle is accelerated since it travels an ever-increasing distance during one swing of the radiofrequency (r.f.) oscillation. At low energy we would expect this distance to increase with the velocity or the root of the kinetic energy but when the energy is large we find that the length of the drift tubes and their spacing no longer increases—a consequence and everyday demonstration of special relativity. The Alvarez structure is still widely used, especially for non-relativistic proton and ion beams.

One may wonder why such a complex structure was used when it was well known at the time that waves might be propagated along a smooth waveguide and that some of the modes have an accelerating electric field in the direction of propagation. We shall see in Chapter 10 that the stumbling block is that the phase velocity of these modes is always greater than that of light and hence the particle sees a field which alternately accelerates and decelerates. It was found only later that the phase velocity could be reduced by a series of iris diaphragms in the pipe. Such a structure (Fig. 1.6) is very popular in electron linacs and also in storage rings when the particle is close to the velocity of light and cavities need not be tuned.

But now we return to the early 1930s to trace the development of yet another kind of accelerator, the cyclotron.

1.5 The cyclotron

Before the linac principle could be fully exploited, another revolutionary idea arrived; that of making a particle follow a circular path in a magnetic field, so that it passes repeatedly through the same accelerating gap. Unlike a linac, whose length must be extended to reach a higher energy, the cyclotron, as it is

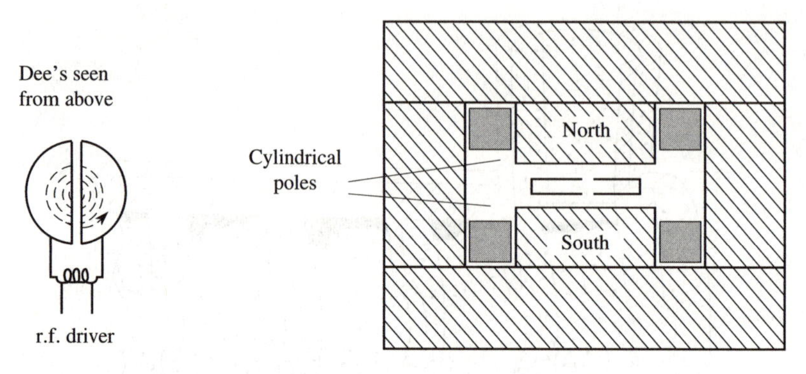

Fig. 1.7 The principle of the cyclotron (from Livingwood 1961).

called, is a relatively compact accelerator in which the energy is only limited by the diameter and field strength of the magnet.

The idea first occurred to E. O. Lawrence (Lawrence and Edelfsen 1930) while he was musing on the possibility of using a magnetic field to recirculate the beam through two of Wideröe's drift tubes. It was published in 1930 and another colleague, M. S. Livingston, who was also later to contribute much to the field, was given the job of making a working model as his doctoral thesis.

In Fig. 1.7 the two 'dee's' can be seen between the poles of the magnet. These, like two halves of a cake-tin sawn along its diameter, are the positive and negative electrodes of the accelerating system. An r.f. generator excites them with an alternating field of constant frequency. The potential difference between the 'dee's' accelerates the ions as they pass the gap between the two halves of the structure. The field oscillates at the particle's circulation frequency and hence the sign of the potential difference at each gap is always in the accelerating direction.

Early cyclotrons were constructed to accelerate ions to modest energies where classical, rather than relativistic mechanics, still applies. In Fig. 1.8, we see the balance between centripetal acceleration of motion in a circle and the force exerted by the vertical magnetic field:

$$evB = \frac{mv^2}{\rho}, \quad \text{if } v \ll c$$

and, rearranging, we obtain the magnetic rigidity—the reluctance of the beam to be bent in a curve:

$$B\rho = \frac{mv}{e}, \quad \text{if } v \ll c.$$

This classical relation may be written in a form that applies also in the relativistic regime if we replace the classical momentum, mv, by the relativistic momentum, p:

$$B\rho = \frac{p}{e}.$$

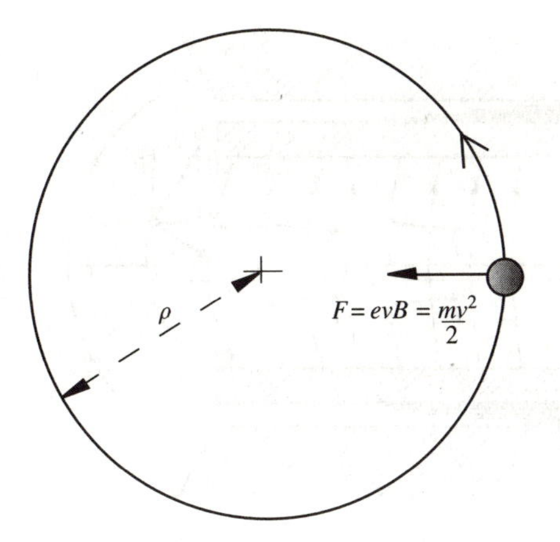

Fig. 1.8 Balance of forces in a cyclotron.

By good fortune the radius of the orbit in a cyclotron is proportional to the velocity and the frequency of revolution:

$$f = \frac{v}{2\pi\rho} = \frac{v}{2\pi} \cdot \frac{eB}{mv} = \frac{eB}{2\pi m}$$

has a numerator and denominator which are both proportional to v. This frequency remains constant as the particle is accelerated. Once the accelerating and revolution frequencies remain in step, a continuous stream of ions injected in the centre will follow a spiral path to reach their highest energy at the rim of the poles. For multiply charged ions e becomes the total charges, q.

1.5.1 Focusing in cyclotrons

When cyclotrons were first developed, very little thought was given to why it was that particles circulated for so many turns. There was clearly a possibility that they might drift away from the mid-plane and hit the pole pieces, but people just accepted this gift of nature and got on with the job of building them. Then one day E. M. McMillan, another of Lawrence's research students and, as we shall see, destined to become famous for other discoveries, was experimenting with ways of changing the radial field distribution by putting a few discs of magnetic material as shims behind the yoke and pole pieces as shown in Fig. 1.9.

It is not clear why he chose to put shims in the middle of the pole, but the effect was dramatic and the cyclotron began to accelerate much higher currents. In retrospect we understand that enhancing the curvature of the fringe field near the pole edge had strengthened horizontal field components which redirect wayward particles heading towards the poles back towards the mid-plane. What is not as graphically obvious is that there is a focusing effect in the horizontal plane too. In order to understand this we must look more rigorously at the dynamics of wayward particles due to a 'centrifugal' focusing effect, but we shall leave this until later. Field gradients are to this day fundamental to accelerator focusing systems.

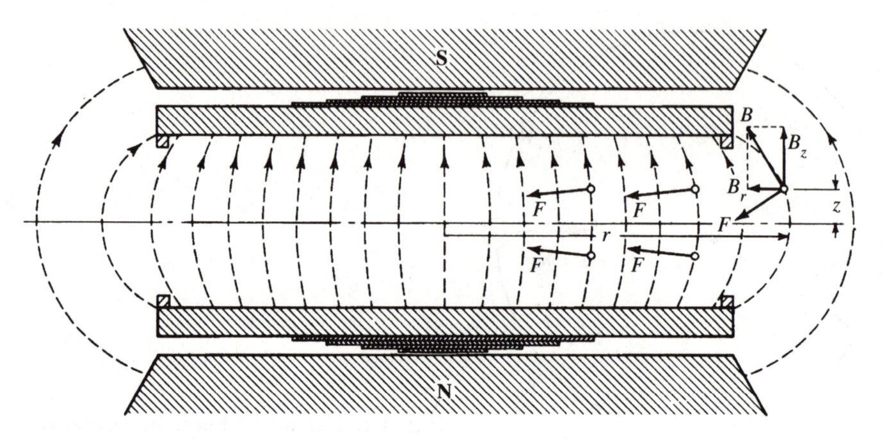

Fig. 1.9 The principle of vertical focusing in a cyclotron (from Livingston and Blewett 1962).

1.5.2 Relativity limits cyclotrons

Lawrence built a series of cyclotrons at Berkeley in the 1930s culminating in a 184 in diameter machine. The sheer size of the poles and return yoke of this machine had reached the practical limit of that time but there was another effect that threatened its successful operation.

Smaller cyclotrons had already experienced difficulty in surpassing the energy of 30 MeV at which the proton begins to become relativistic. It had not occurred to the builders that relativity would become important when the kinetic energy was such a small percentage of the rest energy and it was, therefore, an unpleasant surprise when they discovered their latest and most ambitious cyclotrons were unable to accelerate. Earlier, we saw that the revolution frequency of a cyclotron is constant—an argument that depends upon a cancellation of velocity in numerator and momentum in the denominator and which relies on the classical relation between velocity and momentum. But as momentum and energy continue to increase and the velocity of a particle approaches that of light, its velocity 'saturates'. When the velocity begins to increase less rapidly than the root of the energy, the revolution frequency drops so that particles are no longer synchronous with the accelerating potential.

With the benefit of hindsight, it did not take long for accelerator theorists to find this explanation, but to find and apply a remedy required some ingenuity. In theory, the proton's circulation frequency might be restored by a positive radial field gradient giving a stronger field, which would reduce the radius and the circumference of the orbit at higher energy, but this would destroy the vertical focusing. The alternative is to lower the r.f. frequency to match the sagging revolution frequency. For this to work continuous acceleration would have to be abandoned and the beam injected in regular pulses each following the change in frequency as they are accelerated (Fig. 1.10).

Modifying the frequency of the high-power r.f. generator and maintaining resonance with the system of 'dee's' is not an easy technological challenge but it

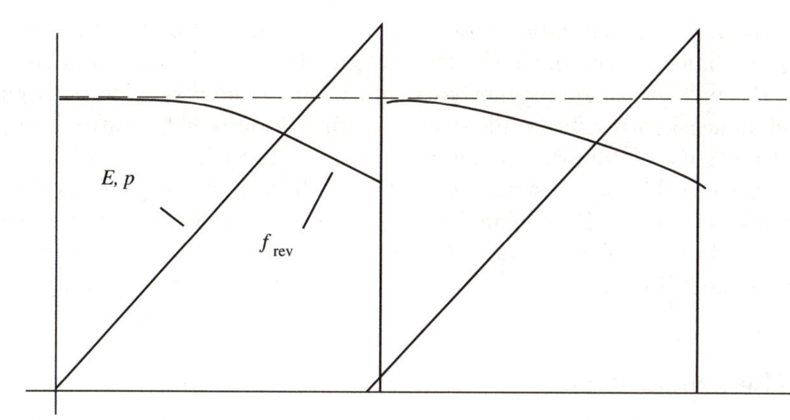

Fig. 1.10 The variation of revolution and r.f. frequencies as a function of time in a synchro-cyclotron.

was solved by tuning the system with a large rotating variable capacitor. In this way the energy of synchro-cyclotrons, as this new version was called, could be extended to many hundred MeV until the sheer mass of steel needed for the poles and their return yokes became prohibitively expensive.

In the early 1940s, cyclotron builders in the US were reassigned to build electromagnetic separators for fissile material and the first attempts to build synchro-cyclotrons had to wait until the end of the war and the discovery of phase stability (see later in this chapter). Lawrence's pre-war 184 in cyclotron was adapted as a synchro-cyclotron and in 1946 accelerated deuterons to 190 MeV and He^{++} to 380 MeV. Other machines followed in the US, in Canada at McGill, in Europe at Harwell in 1949 and at Uppsala, and in Russia at Dubna in 1954.

Later still, thanks to the invention of strong focusing, cyclotron builders found an alternative means to focus the beam rather than decrease the field with radius. They were then able to return to the solution which they had previously rejected of increasing the field radially to compensate the effect of relativity in order to maintain a constant r.f. frequency. In this way cyclotrons again became a continuous source of accelerated particles. Hundreds of such cyclotrons are now used throughout the world mainly for nuclear physics, industrial, and medical applications. But now we must return again to take up the story in the 1940s.

1.6 The betatron

Lawrence's cyclotron programme had been directed to the acceleration of protons and deuterons. To accelerate electrons, which at any useful energy are much too relativistic for cyclotrons, Kerst and Serber reinvented Wideröe's beam transformer idea, renaming it the betatron. Its circular topology was not unlike the cyclotron but it was pulsed and the beam did not spiral out but stayed at the same orbit radius. A short batch of electrons was injected and accelerated by the rate of change of the magnetic flux linking the orbit. Kerst found that by shaping the poles, a single magnet yoke and winding could provide both guide field and an accelerating flux while still respecting Wideröe's 2 : 1 ratio. Working at the University of Illinois and later at General Electric Corporation Laboratories in the USA, he constructed machines which surpassed the energies of Lawrence's cyclotrons. By the mid-1940s betatrons had begun to become as bulky as cyclotrons. The magnet of a 300 MeV machine at the University of Illinois weighed 275 ton.

1.7 The synchrotron

After the war, still higher energies were needed to pursue the aims of physics and the stage was set for the discovery of the synchrotron principle opening the way to the series of circular accelerators and storage rings which have served particle physics up to the present day. It was Australian physicist Mark Oliphant, then supervising uranium separation at Oak Ridge, who synthesized three old ideas into a new concept—the synchrotron. The ideas were: accelerating between the gaps of resonators, varying the frequency, and pulsing the magnet. In 1943, he described his invention in a memo to the UK Atomic Energy Directorate:

> 'Particles should be constrained to move in a circle of constant radius thus enabling the use of an annular ring of magnetic field ... which would be varied in such a way that the radius of curvature remains constant as the particles gain energy through successive accelerations by an alternating electric field applied between coaxial hollow electrodes.'

We see in Fig. 1.11 how, once a short pulse is injected at low field, the field rises in proportion to the momentum of particles as they are accelerated and this ensures that the radius of the orbit remains constant. Unlike cyclotrons and betatrons, the synchrotron needs no massive poles to support a magnetic field

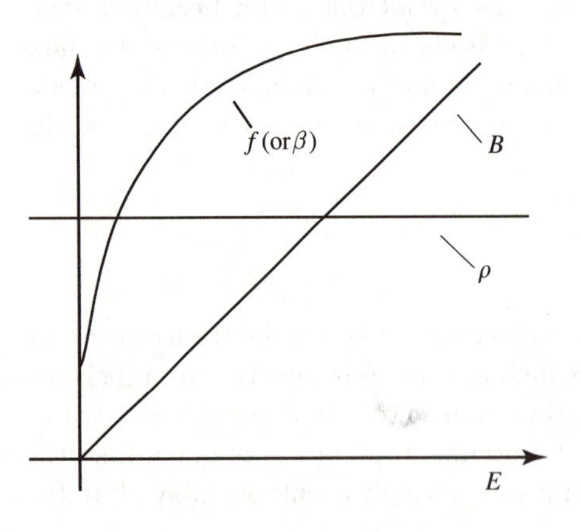

Fig. 1.11 Field and frequency rise together in a synchrotron.

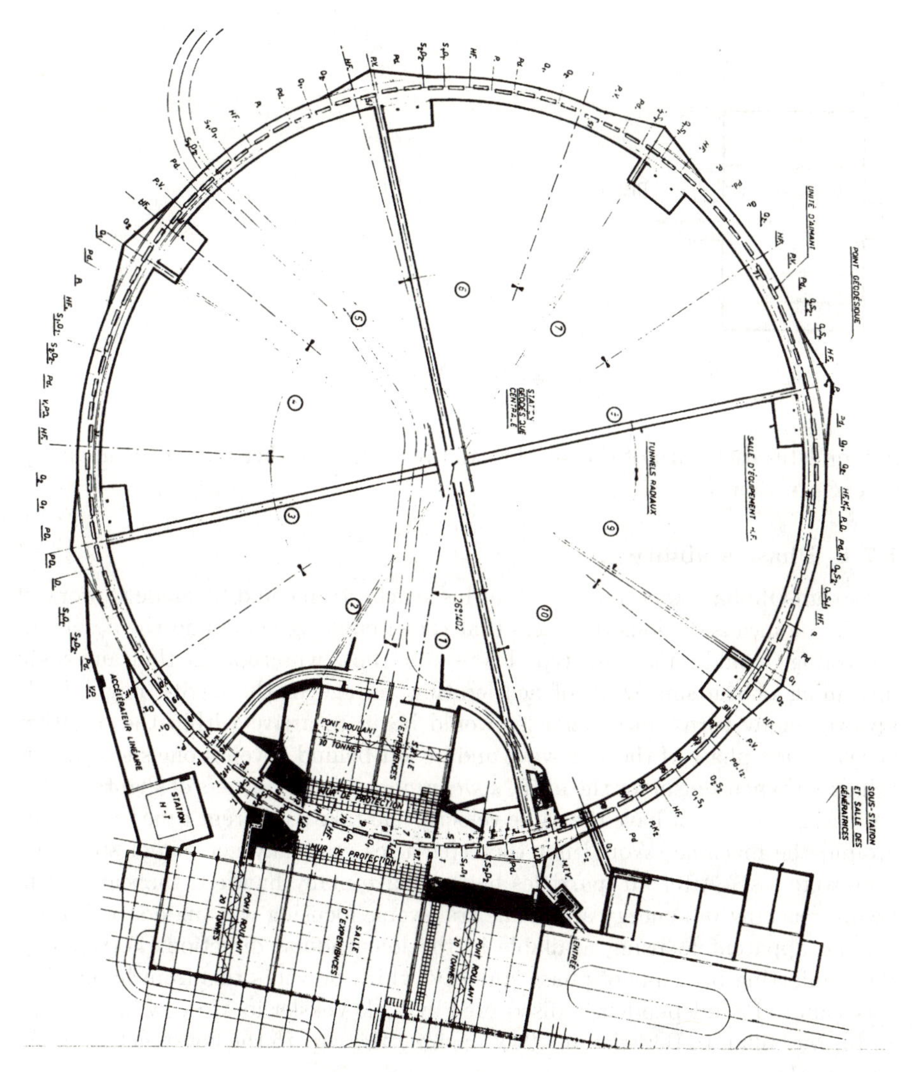

Fig. 1.12 The CERN 25 GeV proton synchrotron.

within the beam's circular orbit. The guide field is instead provided by a slender ring of individual magnets (see for example Fig. 1.12). The fact that the machine is pulsed and the frequency must be controlled to track the increasing speed of particles is a complication, but it solves the difficulty that isochronous cyclotron builders had encountered in accelerating relativistic particles. Incidentally, the flux linking the orbit is much too weak to provide any betatron acceleration.

Acceleration is provided by fields within a hollow cylindrical resonator, Fig. 1.13, excited by a radio transmitter. A particle passes from left to right as

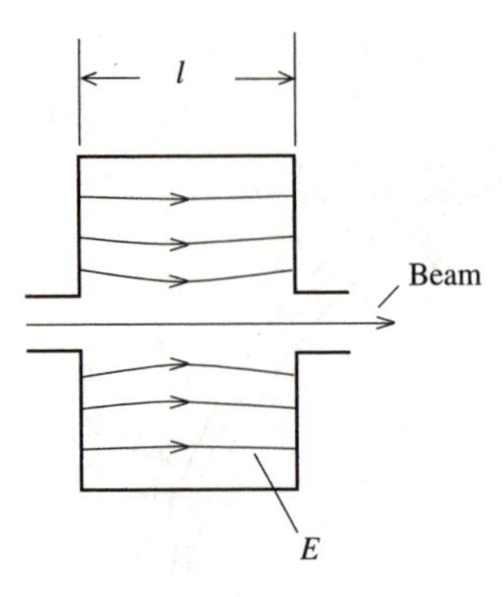

Fig. 1.13 A simple accelerating cavity.

it completes each turn of the synchrotron receiving another increment in energy on each revolution.

1.7.1 Phase stability

Although Oliphant was confident that his synchrotron could be made to work, it was by no means obvious to others that the circulating beam and the accelerating voltage would remain in step. There were those who thought that any slight mistiming of the sine wave of accelerating voltage in the cavity might build up over many turns until particles would begin to arrive within the negative, decelerating, phase of the sine wave and be left behind. Even if one succeeded in achieving synchronism for the ideal, *synchronous particle*, others of slightly different energy would not have the same velocity and take a different time to circulate around the machine. Would not these particles gradually get out of step until they were lost? After all, particles had to make many hundred thousand turns before reaching full energy and while transverse focusing was understood, there was no apparent focusing available in the longitudinal direction. Fortunately, the comforting principle of phase stability, which prevents this from happening, was soon to be independently discovered by V. I. Veksler in Moscow in 1944 and E. M. McMillan in Berkeley in 1945, opening the way to the construction of the first synchrotrons.

In order to achieve phase stability, particles orbiting the synchrotron are timed to ride, not on the peak, but on the flank of the voltage wave in the accelerating cavity. They receive more or less energy than the synchronous particle so that they oscillate about the stable or synchronous phase. For all particles, the time average of their energy gain matches the rising magnetic field. We shall come to discuss this principle of phase stability in Chapter 5.

1.7.2 The first synchrotrons

Before Oliphant's synchrotron proposal had emerged from the classified files of war-time Britain, McMillan had scooped the idea and written a letter to the editor of *Physical Review* announcing, not only the discovery of phase stability, but proposing a 'synchrotron' defined as a machine in which both frequency and magnetic field vary. McMillan was promptly given the green light to construct an electron synchrotron of 300 MeV at the University of California, but there were other enthusiasts in the field eager to be the first to prove the principle.

At the Telecommunications Research Laboratory in Malvern UK, Frank Goward and his colleague D. E. Barnes, hearing of McMillan's work on the synchrotron and phase stability, modified a small betatron to operate as a synchrotron (Goward and Barnes 1946). By adding accelerating electrodes they were able to prolong acceleration beyond the limit at which the 2 : 1 flux ratio broke down because of saturation, to reach roughly twice the energy. This first proof of the synchrotron principle took place in August 1946 just ahead of a team at the General Electric Co. at Schenectady who were constructing a purpose-built 70 MeV electron synchrotron (Elder *et al.* 1947). This machine, which had a glass vacuum chamber, had the distinction of being the first to produce visible synchrotron radiation, a phenomenon we shall come to discuss later.

These early synchrotrons were electron machines but projects for proton synchrotrons aiming at energies above 1 GeV were not far behind. Oliphant, now back at the University of Birmingham, had been the first to start a proton synchrotron (1 GeV) but lack of funds and graduate labour delayed completion (Oliphant 1967).

Meanwhile, in 1948, construction started on two huge proton synchrotrons. The Bevatron, aimed at 6 GeV, or 6 billion electron volts in the US parlance, was started at the University of California, Berkeley while on the East Coast, the Brookhaven National Laboratory set about the 3 GeV Cosmotron. It was the Cosmotron team who won the race and in May 1952 the *New York Times* headlined their first 'Billion Volt Shot' (Fig. 1.14).

1.7.3 Weak focusing

Cyclotron builders had discovered that the beam could be prevented from hitting the upper and lower pole pieces by adding vertical transverse focusing. A field with a gradient in the range:

$$0 < n = -\frac{r}{B}\frac{\partial B_z}{\partial r} < 1$$

was strong enough to provide vertical focusing but its inevitable defocusing effect in the radial plane would not swamp the natural radial focusing from centrifugal forces. Early synchrotron builders exploited this principle and magnet poles were shaped to provide a constant gradient decreasing towards the outside. Such focusing was weak and the excursions of the beam large (Livingston and Blewett 1962). Pole widths and gaps in some constant gradient machines were large enough for people to crawl through (Judd private communication).

Fig. 1.14 The Cosmotron (photo: Brookhaven National Laboratory).

1.7.4 Strong focusing

Strong focusing changed this. It was invented at the Cosmotron whose weak focusing 'C'-shaped magnet was open to the outside. The top energy of the Cosmotron was limited by the extra fall-off in field caused by the effect of saturation. Eventually, as saturation set in, n became greater than unity and cancelled horizontal focusing. E. D. Courant, S. Livingston and H. S. Snyder wanted to compensate this by re-installing some of the C-magnets with their return yokes towards the outside. They were afraid of the variations in gradient around the ring but were surprised to calculate that the focusing seemed to improve as the strength of the alternating component of the gradient increased. Courant and Snyder (1958) were able to explain this retrospectively with an optical analogy of alternating focusing by equal convex and concave lenses which will transport rays which pass through the centres of defocusing lenses.

They found to their disappointment that this idea had actually been patented earlier by one of their colleagues Christofilos (1950). Alternating-gradient or strong focusing, greatly reduces the beam's excursions and the cross section of the magnet gap can be reduced to become comparable with a hand rather than a whole human body. Its discovery enabled Brookhaven and CERN to build the next generation of proton synchrotrons, the Alternating Gradient Synchrotron, AGS, and the CERN Proton Synchrotron, CPS, to reach 30 GeV—five times the energy of the Bevatron—yet use beam pipes of only a few centimetres height and width.

1.7.5 Fixed field alternating gradient

There were further ripples created by the strong focusing discovery. Those working on cyclotrons in the Midwestern Universities Research Association (MURA), centred in Chicago, realized that a gradually increasing radial field might be combined with strong focusing to resemble a ridged polepiece idea proposed much earlier by L. H. Thomas. Thomas had invented this alternative to the synchrocyclotron just before the war but his paper had been ignored as too difficult to understand. Alternating-gradient focusing is so strong that the cyclotron's guide field can *increase* radially so that particles can remain in synchronism as they are accelerated into the relativistic regime. The defocusing due to the positive gradient is small in comparison to the alternating-gradient focusing. This cut the Gordian knot that had meanwhile forced cyclotron designers to resort to pulsed operation. The cyclotron again became a continuous source of particles.

By making the radial field gradient very strong and sweeping the ridges into a spiral they proposed a fixed field alternating-gradient (FFAG) accelerator. The r.f. frequency is varied to accelerate from injection to top energy within a narrow band of radius without pulsing the magnet thus rivalling the pulsed synchrotron.

The study of the non-linear fields in FFAG machines laid the foundations of the theory of field errors and stimulated tracking of particles by computer simulation. MURA also studied tracking in the longitudinal direction as the beam is accelerated. With this came the important realization that beams could be accumulated side by side by phase displacement of the r.f.—a process known as stacking. MURA workers and those at Novosibirsk had also to wrestle with understanding the innumerable instabilities which threaten intense beams. This work proved invaluable preparation for the construction of the accelerators and colliders that were to follow.

Although the FFAG concept, eventually, gave birth to the modern sector-focused cyclotron, it came rather too late to influence the plans to build powerful synchrotrons in the US and at CERN which had already become firmly rooted in their own version of strong focusing. However, one idea mentioned by Kerst *et al.* (1956) when they published the FFAG concept was taken up eagerly by others. They had pointed out that by joining two such machines in a figure of eight, beams of particles might collide head-on. We mentioned earlier that such head-on collisions are much more effective as much of the energy carried by a particle in collision with a fixed target is 'wasted', taken away as kinetic energy by the interaction products. It was not a new idea and had apparently featured among physics examination questions in the 1930s, but Kerst's paper certainly served as a trigger to bring it to the attention of accelerator builders in the US.

The advantage of head-on collisions had occurred even earlier to Wideröe but it was Kerst who persuaded B. Touschek to construct the first storage ring in which a beam of electrons, once accelerated, could circulate indefinitely colliding with a counter rotating beam of positrons. Such storage rings and colliders have dominated the recent use of particle accelerators for high-energy physics and we shall chart their history in detail in Chapter 11. However, we should mention the next large CERN project—the Intersecting Storage Rings (ISR) (Johnsen 1964). The energy of these beams—30 GeV—provided a huge leap in centre of mass energy for physics and served as a test bed for the first large all-purpose particle detectors. The ISR was followed by two, much larger fixed-target proton accelerators. The first, completed in 1971, was the 400 GeV (later 500 GeV) FNAL synchrotron with a circumference of more than 6 km (Wilson, R. R. 1971). This was followed, five years later, by a similar machine, the Super Proton Synchrotron (SPS) at CERN (Wilson, E. J. N. (ed.) 1972).

1.8 Superconducting magnets

The radius of a synchrotron is governed by the magnetic rigidity of the beam which, as in a cyclotron, is proportional to its top energy (strictly its momentum) and inversely to its magnetic field. As synchrotrons and proton colliders have grown, their builders have sought stronger magnetic fields to reduce the real estate they occupy and the cost of excavating the ring tunnel. Room-temperature magnets have steel pole pieces which define the field shape which, together with the return yoke, usually saturate at a field of about 2 T. One can, of course, imagine air cored windings precisely shaped around a cylindrical pipe to produce a uniform field without the need for iron. The ideal coil shape should mimic a pair of intersecting ellipses as closely as possible. However, the field in such a magnet is severely limited by the problem of cooling the coils which must carry a considerable current density to generate a useful continuous field. Superconducting coils, which in theory do not dissipate heat, offer a means to increase the current density and have allowed modern synchrotron designers to exploit this geometry and increase the guide field to 4 T and recently, 8 T.

The first large machine to exploit superconducting technology was the Tevatron—a superconducting ring nestling between the supports of Fermilab's 6 km circumference main ring (Griffin 1980). The Tevatron extended the energy of the facility to 1000 GeV and later became a proton–antiproton collider emulating the success of CERN's Sp$\bar{p}$S which in the early 1980s had pioneered colliding beam physics with 350 GeV hadron beams.

In Europe, came DESY's HERA which stored 820 GeV protons in a superconducting ring almost as big as the Tevatron and collided them with 30 GeV electrons, and of course LEP, the 50 on 50 (later to become 100 on 100) GeV electron positron collider at CERN. This machine, building on experience of similar 30 GeV rings at SLAC and DESY, must surely be the ultimate size for circular lepton colliders (CERN 1984). Higher energy electron rings would emit too much synchrotron radiation.

LEP has now reached the end of its useful life and once again the pendulum will swing back from lepton to hadron accelerators at CERN. LEP will be dismantled and in its place a Large Hadron Collider (LHC) installed. Exploiting the 27 km LEP tunnel and using the most advanced superconducting magnet technology, this machine will collide two 7 TeV proton beams circulating in its twin-bore superconducting magnets (Lefèvre 1995). The intersecting geometry is not unlike the ISR and once again shortage of antiprotons has prompted its builders to opt for two proton beams. The magnetic field of 8 T is made possible by cooling the superconductor down to 1.8 K where the liquid helium coolant is superfluid and an almost perfect thermal conductor. For the first time, synchrotron radiation from a proton beam must be taken into account in assessing the heat load to the magnet and its cryostat—a factor which, though not critical for LHC, may limit the luminosity of any larger circular hadron colliders that may follow. Another use of the LHC will be to collide beams of heavy ions following up the studies at RHIC, the Relativistic Heavy Ion Collider now starting at the Brookhaven National Laboratory, USA.

One may speculate on the machine to follow LHC—perhaps a larger hadron storage ring, a linear lepton collider or even a muon storage ring. Readers should make up their own minds after reading Chapters 11 and 14!

1.9 Accelerators at work in medicine, industry, and research

We have reviewed a history of accelerators which has been largely driven by the needs of particle physics. We should not forget, however, that many thousands of accelerators have been put to more practical use in other branches of scientific research as well as in industry and medicine.

Small linear accelerators and betatrons are commonplace tools for cancer therapy in the more advanced medical centres of the world. Many cyclotrons are at work producing isotopes which can be used as radioactive tracers in industry as well as isotopes which emit positrons and which can be attached to biochemical molecules used in medicine. Computer analysis of the pattern of positron emission coupled with body scans are a powerful diagnostic technique.

Recently, proton synchrotrons of a few hundred MeV have been built in the US and in Japan to irradiate deep tumours. In Europe, the PSI cyclotron as well as the ion accelerators at GSI, Darmstadt, have been put to this use and there are plans to build a dedicated synchrotron facility in Italy.

In industry, electron beams of quite low energy are used to cure paint coatings, polymerize plastics and to sterilize medical supplies and even foodstuffs. Heavy-ion beams, such as those accelerated by GSI, are widely used to implant atoms in the surfaces of semiconductors to 'print' the circuits of modern computer chips. Other industrial uses are hardening metal surfaces for bearings and etching silicon microcircuitry.

Recent years have witnessed a mushrooming growth of synchrotron radiation sources all over the world and their highly collimated and tuneable radiation is put to a huge variety of applications. In research, X-ray diffraction techniques reveal the structure of proteins and enzymes and the crystal lattices of exciting new materials such as high-temperature superconductors. These applications of synchrotron radiation are complemented by neutrons from intense proton machines—the so-called spallation sources—such as ISIS at RAL. Neutron diffraction extends many of the research techniques of synchrotron light sources and adds a new dimension in that the source may be pulsed to allow time of flight identification techniques.

Even more impressive accelerators of only 1 GeV or so, but designed for very high current, are under study to bring pellets of deuteron/tritium into the conditions required for a self-sustained thermonuclear reaction. Intense linear accelerators or cyclotrons would also allow transmutation of long-lived nuclear waste into isotopes which rapidly decay to become harmless or alternatively provide the beam which 'fans the flames' of the 'energy amplifier'—a fail-safe form of nuclear reactor using relatively innocuous thorium fuel.

1.10 Moving from history to physics

This chapter was intended to provide an overview of the main features which distinguish the different kinds of accelerators. It is now time to examine in a more mathematical way the fundamental principles of these machines.

Exercises

1.1 In special relativity the rest energy of the particle is defined $E_0 = m_0c^2$, where c is the velocity of light and m_0 is the rest mass. Write down the expressions for the total energy E and the momentum p in terms of E_0 and kinetic energy, T.

1.2 Using the notation of special relativity $\beta = v/c$ and $\gamma = 1/\sqrt{1-\beta^2}$ show that $\gamma = E/E_0$ and $\beta = pc/E$.

1.3 The kinetic energy T of a proton is 1 GeV. If its rest mass m is 0.9383 GeV/c^2, what is its total energy?

1.4 Given that the relation between relativistic momentum and total energy is

$$E^2 = (m_0c^2) + (pc)^2,$$

calculate its momentum (in GeV/c).

1.5 A betatron has a beam radius of 0.1 m and is powered from 50 Hz mains. Its peak guide field is 1 T while the flux linking the orbit is twice that which would result from a uniform field of this value. What will be the peak energy of the electrons it accelerates?

1.6 Using classical mechanics show that the angular frequency of revolution of a proton in a cyclotron is equal to $B_z(e/m)$. Calculate this frequency for a field of 1–2 T ($e/m = 9.58 \times 10^7$ C/kg).

1.7 A synchrotron of 25 m radius accelerates protons from a kinetic energy of 50 to 1000 MeV in 1 s. The dipole magnets saturate at 1000 MeV. What is the maximum energy of deuteron ($Z = 1$, $A = 2$) that it could accelerate? Hint: for protons use the expression derived in Section 2.2:

$$B\rho = \frac{p}{e}.$$

1.8 What is the revolution frequency for (a) protons and (b) deuterons?

2

Transverse motion

2.1 Description of motion

2.1.1 Coordinate system

The bending fields of a synchrotron are usually vertically directed, causing the particle to follow a curved path in the horizontal plane (Fig. 2.1). The force acting on the particle is horizontal and is given by

$$\mathbf{F} = e\mathbf{v} \times \mathbf{B},$$

where $\mathbf{v}$ is the velocity of the charged particle in the direction tangential to its path and $\mathbf{B}$ is the magnetic guide field.

If the guide field is uniform, the ideal motion of the particle is simply a circle of radius of curvature ρ, but we can also define a local radius of curvature $\rho(s)$ to describe motion in a non-uniform field. We shall suppose that it is possible to find an orbit or curved path for the particle which closes on itself around the synchrotron, which we call the equilibrium orbit. The machine is usually designed with this orbit at the centre of its vacuum chamber.

2.1.2 Displacement and divergence

Of course a beam of particles enters the machine as a bundle of trajectories spread about the ideal orbit. At any instant a particle may be displaced horizontally

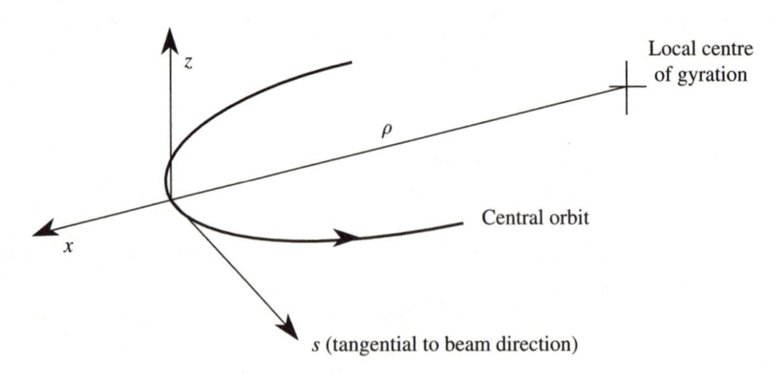

Fig. 2.1 Charged particle orbit in magnetic field.

by x and vertically by z from the ideal position and may also have divergence angles horizontally and vertically with respect to the central orbit:

$$x' = \frac{dx}{ds} \quad \text{and} \quad z' = \frac{dz}{ds}.$$

Such mis-steering would cause particles to leave the vacuum pipe were it not for the carefully shaped field which restores them back towards the beam centre so that they oscillate about the ideal orbit. The design of the restoring fields determines the transverse excursions of the beam and the size of the cross section of the magnets and is therefore of crucial importance to the cost of a project.

2.1.3 The betatron envelopes

A modern synchrotron consists of pure bending magnets and quadrupole magnets or lenses which provide focusing. These are interspersed among the bending magnets of the ring in a pattern called the lattice. In Fig. 2.2 we see an example of such a magnet pattern which is one cell, or about 1% of the circumference, of the 400 GeV SPS at CERN. Although the SPS is now considered a rather old-fashioned machine, its simplicity leads us to use it frequently as an example in this book. This focusing structure is called FODO which describes the sequence of quadrupoles which focus or defocus the beam. The envelope of these oscillations follows a function $\beta(s)$ which has waists near each defocusing magnet and has a

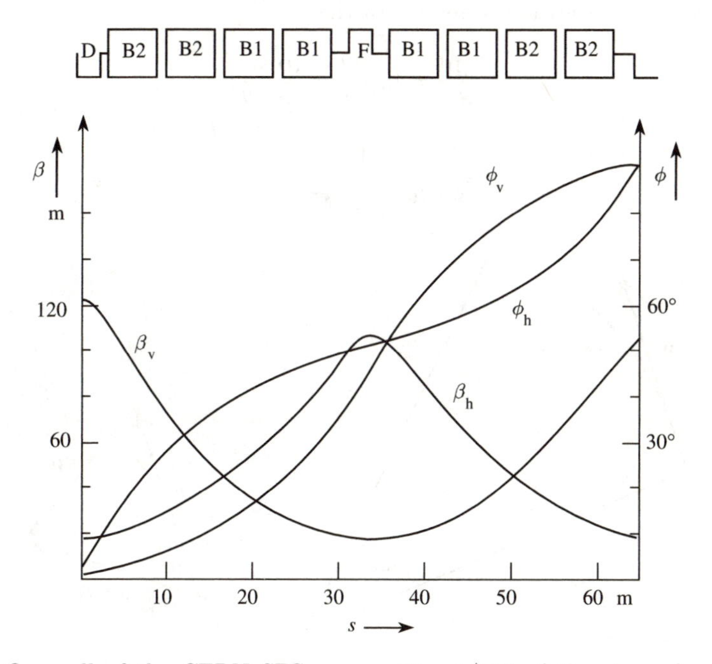

Fig. 2.2 One cell of the CERN SPS representing 1/108 of the circumference. The pattern of dipole (B) magnets and quadrupole (F and D) lenses is shown above.

maximum at the centres of F quadrupoles. Since F quadrupoles in the horizontal plane are D quadrupoles vertically, and vice versa, the two functions $\beta_{\mathrm{h}}(s)$ and $\beta_{\mathrm{v}}(s)$ are one half-cell out of register in the two transverse planes. The function β has the dimensions of length but the units bear no direct relation to the physical beam size. The reader should be clear that particles do not follow the $\beta(s)$ curves but oscillate within them in a form of modified sinusoidal motion whose phase advance is described by $\phi(s)$. The phase change per cell, in the example shown, is close to $\pi/2$ but the rate of phase advance is modulated throughout the cell.

We will develop the ideas behind this description in the following sections; but let us first establish the concepts of the magnets which bend the beam in a circle.

2.2 Bending magnets and magnetic rigidity

Suppose the particle has a relativistic momentum vector $\mathbf{p}$ and travels perpendicular to a field $\mathbf{B}$, which is into the plane of the diagram (Fig. 2.3). After time dt it has followed a curved path of radius ρ whose length is ds and its new momentum is $\mathbf{p}+d\mathbf{p}$. We may equate the force and rate of change of momentum:

$$e\mathbf{v}\times\mathbf{B}=\frac{d\mathbf{p}}{dt},$$

and we see from resolution of momenta that

$$\frac{d|\mathbf{p}|}{dt}=|\mathbf{p}|\frac{d\theta}{dt}=\frac{|\mathbf{p}|}{\rho}\frac{ds}{dt}.$$

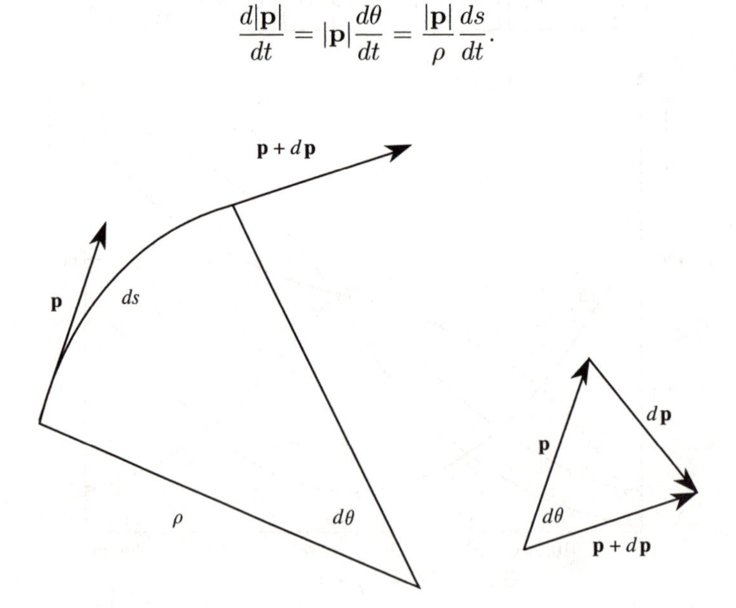

Fig. 2.3 Vector diagram showing differential changes in momentum for a particle trajectory.

On the other hand, if the field and plane of motion are normal, the magnitude of the force may be written:

$$e|\mathbf{v} \times \mathbf{B}| = e|\mathbf{B}|\frac{ds}{dt}.$$

Equating the right-hand sides of the two expressions above, we find that we can define the quantity known as magnetic rigidity:

$$B\rho = \frac{p}{e}.$$

Strictly, we should use the units N s for p and express e in coulombs to give $B\rho$ in T m. However, in charged particle dynamics we often talk about the 'momentum' pc which has the dimensions of an energy and is expressed in units of GeV. A useful rule of thumb formula based on these units is

$$B\rho\ (\mathrm{T\,m}) = 3.3356p\ \ (\mathrm{GeV}/c).$$

Figure 2.4 shows the trajectory of a particle in a bending magnet or dipole of length l. Usually, the magnet is placed symmetrically about the arc of the particle's path. One may see from the geometry that

$$\sin\frac{\theta}{2} = \frac{l}{2\rho} = \frac{lB}{2(B\rho)},$$

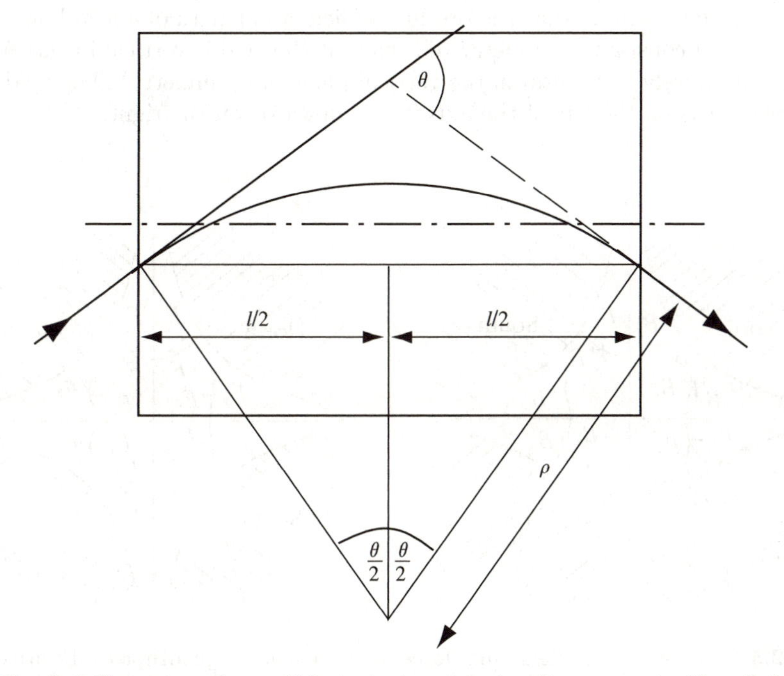

Fig. 2.4 Geometry of a particle trajectory in a bending magnet of length l.

and, if $\theta \ll \pi/2$,

$$\theta \approx \frac{lB}{B\rho}.$$

The bending magnet aperture must be wide enough to contain the sagitta of the beam, which is the distance between the apex of the arc and the chord:

$$\pm\rho\left(1 - \cos\frac{\theta}{2}\right) \approx \pm\frac{\rho\theta^2}{8} \approx \frac{l\theta}{8}.$$

The ends of bending magnets are often parallel but in some machines are designed to be normal to the beam. There is a focusing effect at the end which depends on the angle of these faces. We will come back to this in the next chapter.

2.3 Focusing

2.3.1 Quadrupole magnets

The principal focusing elements in modern synchrotrons are quadrupole magnets. The poles are truncated rectangular hyperbolae and alternate in polarity. Figure 2.5 shows a particle's view of the fields and forces in the aperture of a quadrupole as it passes through normal to the plane of the paper. The field shape is such that it is zero on the axis of the device but its strength rises linearly with distance from the axis. This can be seen from a superficial examination of Fig. 2.5 if we remember that the product of field and length of a field line joining the poles is a constant. Symmetry tells us that the field is vertical in the median plane (and purely horizontal in the vertical plane of asymmetry). The field must be downwards on the left of the axis if it is upwards on the right.

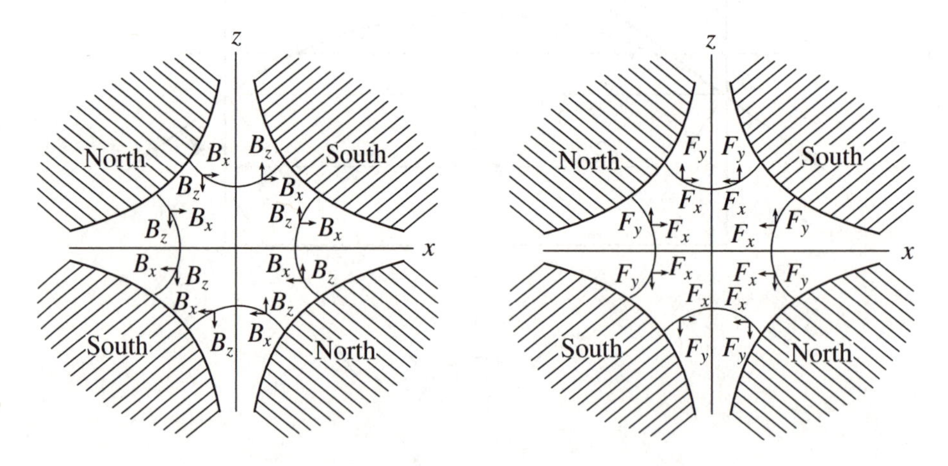

Fig. 2.5 Components of field and force in a magnetic quadrupole. Positive ions approach the reader on paths parallel to the s-axis (Livingood 1961).

This last observation ensures that the horizontal focusing force, evB_z, has an inward direction on both sides and, like the restoring force of a spring, rises linearly with displacement, x. The strength of the quadrupole is characterized by its gradient dB_z/dx normalized with respect to magnetic rigidity:

$$k = \frac{1}{B\rho}\frac{dB_z}{dx}.$$

The angular deflection given to a particle passing through a short quadrupole of length l and strength k at a displacement x is therefore:

$$\Delta x' = \theta = \frac{lB}{B\rho} = \frac{l(dB_z/dx)x}{B\rho} = lkx.$$

The use of x' to indicate the divergence angle of a trajectory is defined in Fig. 2.7. Compare this with a converging lens in optics:

$$\Delta x' = -\frac{x}{f}$$

and we see that the focal length of a horizontally focusing quadrupole is

$$f = -\frac{1}{kl}.$$

The particular quadrupole shown in Fig. 2.5 would focus positive particles coming out of the paper or negative particles going into the paper in the horizontal plane. A closer examination reveals that such a quadrupole deflects particles with a vertical displacement away from the axis—vertical displacements are defocused. This can be seen if Fig. 2.5 is rotated through 90°.

2.3.2 The gutter analogy

It is important to start with a tangible concept of focusing and so we digress for a moment to consider a focusing system which is much simpler. Let us ignore vertical defocusing for a moment and consider the horizontal focusing of an infinitely long quadrupole, this is in fact exactly the focusing system of the weak-focusing synchrotron discussed in Chapter 1. A particle oscillates in this focusing system like a small sphere rolling down a slightly inclined gutter with constant speed. Figure 2.6 shows two views of this motion and from the right hand we recognize the motion as a sine wave. Note too that the sphere makes four complete oscillations along the gutter. In the language of accelerators, its motion has a wavenumber $Q = 4$.

Now let us extend this analogy by bending the gutter into a circle rather like the brim of a hat. We provide the necessary instrumentation to measure the displacement of the sphere from the centre of the gutter each time it passes a given mark on the brim and we also have a means to measure its transverse velocity. With the aid of a computer, we might convert this information into the

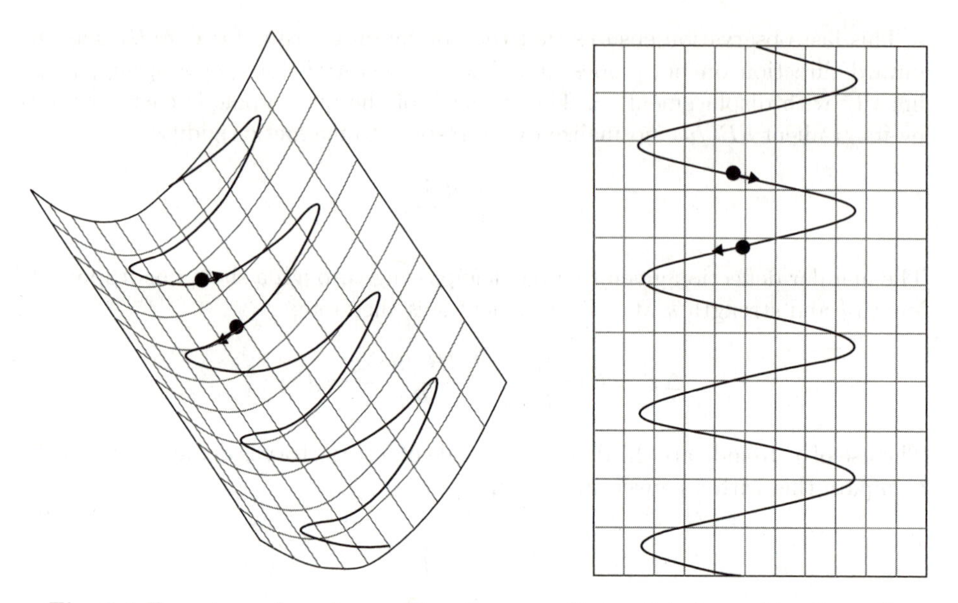

Fig. 2.6 Two views of a sphere rolling down a gutter as it is focused by the walls.

divergence angle shown in Fig. 2.7:

$$x' = \frac{dx}{ds} = \frac{v_\perp}{v_\parallel}.$$

Suppose also that we make the brim of a hat out of a slightly different length of gutter than is shown so that Q is not an integer. We can plot a point for each arrival of the sphere in a diagram of x' against x, which we call a 'phase-space diagram' of transverse motion. The sphere has a large transverse velocity as it crosses the axis of the gutter and has almost zero transverse velocity as it reaches its maximum displacement.

The locus of the 'observations' will be an ellipse (Fig. 2.7) and the phase will advance by Q revolutions each time the particle returns. Of course, only the fractional part of Q may be deduced from our observations since we are blind to what happens round the rest of the hat's brim—a situation we shall find is common in the real life of accelerators.

In order to establish concepts which will take us from the gutter analogy to real synchrotrons, we have to define some of the transverse beam dynamical quantities more rigorously. The area of the ellipse is a measure of how much the particle departs from the ideal trajectory which, in the diagram, is represented by the origin:

$$\text{Area} = \pi\varepsilon \ \ (\text{mm rad}).$$

In accelerator notation we use ε, the product of the semi-axes of the ellipse, as a measure of the area called the emittance. The emittance is usually quoted in

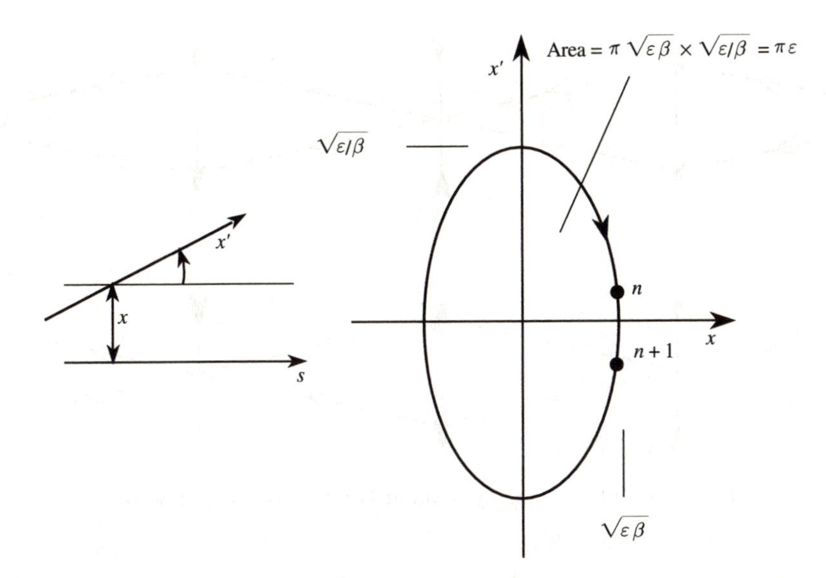

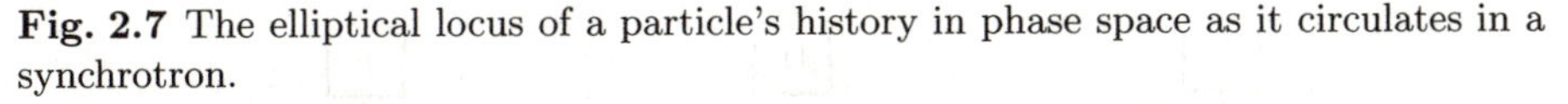

Fig. 2.7 The elliptical locus of a particle's history in phase space as it circulates in a synchrotron.

units of π mm mrad. The maximum excursion in displacement, the major axis, of the ellipse is defined as

$$\hat{x} = \sqrt{\varepsilon\beta},$$

hence,

$$\hat{x}' = \sqrt{\frac{\varepsilon}{\beta}}.$$

The quantity β is a property of the gutter, not the beam. In the synchrotron it varies around the ring and is the envelope function plotted in Fig. 2.2 and again in Fig. 2.9. By analogy, the brim of the hat, which represents the alternating gradient focusing system shown in this figure, will vary its width and curvature around the crown and β will follow this variation in some way. Note that the aspect ratio of the ellipse is just β. We will return to these quantities when we have studied more about the alternating-gradient focusing systems.

2.3.3 Alternating-gradient focusing

We explained in Chapter 1 how the discovery of alternating-gradient focusing (Courant and Snyder 1958) was a major breakthrough in the design of synchrotrons which allowed designers to use quadrupoles in spite of their defocusing property in one plane. It enables much stronger focusing systems to be used with considerable savings in the space needed for the beam cross section.

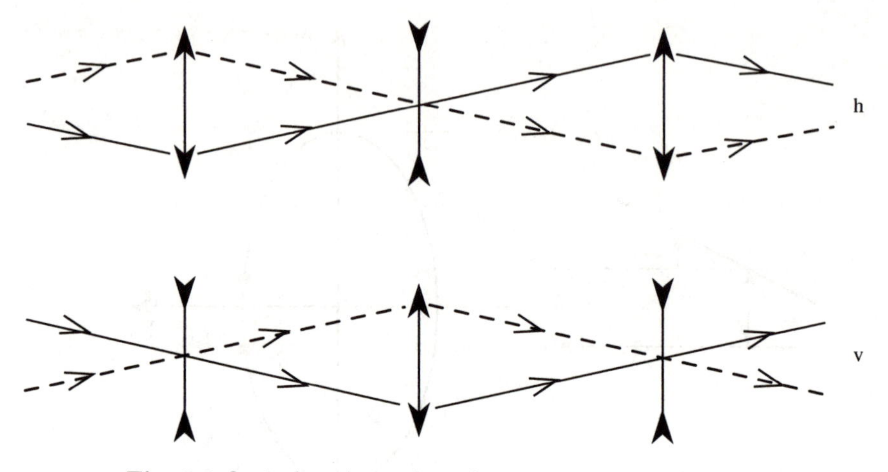

Fig. 2.8 Optical analogy of an alternating pattern of lenses.

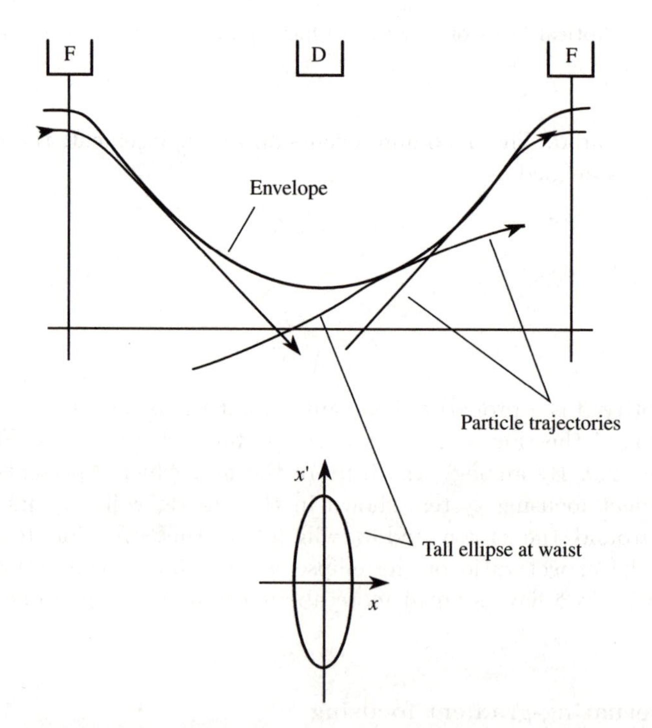

Fig. 2.9 The paths of particles within a FODO lattice are within the envelope of betatron motion and, like the rays of Fig. 2.8, are always closer in the D quadrupoles so they receive a net focusing effect. The phase-space ellipse is tall and narrow at the D lens where the beam has a large divergence spread.

The principle is illustrated in Fig. 2.8, which shows an optical system in which each lens is concave in one plane while convex in the other. It is possible, even with lenses of equal strength, to find a ray which is always on axis at the D lenses in the horizontal plane and therefore sees only the F lenses. The spacing of the lenses would then have to be $2f$. If the ray is also central in the lenses which are vertically defocusing, the same condition will apply simultaneously in the vertical plane. At least one particular particle or trajectory corresponding to this ray will be contained indefinitely.

The alternating-gradient idea will work even when the rays in the D lenses do not pass dead centre and the lenses are not spaced by exactly $2f$. In fact, it is sufficient for the particle trajectories to *tend* to be closer to the axis in D lenses than in F lenses as shown in Fig. 2.9.

By suitable choice of strength and spacing of the lenses the envelope function $\beta(s)$ can be made periodic in such a way that it is large at all F quadrupoles and small at all D's. Symmetry will ensure that this is true also in the vertical plane. Particles oscillating within this envelope will always tend to be further away from the axis in F quadrupoles than in D quadrupoles and there will, therefore, be a net focusing action. We have already seen that β is the aspect ratio of the phase-space ellipse. At F quadrupoles the ellipse will be squat and at D quadrupoles it will be tall. In the next chapter we shall define this envelope or betatron amplitude more rigorously and establish how to calculate it for a given lattice of focusing magnets (see also Schmüser 1987; Rossbach and Schmüser 1992).

Exercises

2.1 The radius of the old ISR tunnel at CERN is 150 m. We wish to design a new electron ring with this radius. Only 43% of the circumference can be bending magnets; what will be the local radius of bend ρ in these magnets if they all have the same strength?

2.2 The momentum of the electrons is to be 8 GeV/c. What is the $B\rho$? From this calculate the field in the dipoles.

2.3 The bore of the quadrupoles is 70 mm and the field at this radius is 0.42 T; what is the maximum field gradient?

2.4 There are 144 quadrupoles and the focal length of each should be about 2/3 of their spacing. How long should they be? (See Fig. 2.2.)

2.5 We allow 0.5 m at one end of a dipole and 1.5 m at the other and we alternate dipoles and quadrupoles. How long must the dipoles be?

2.6 How many bending magnets will there be and how many empty half-periods?

2.7 Now recalculate the bending radius and the filling factor.

3 Lattices

3.1 Betatron envelopes

During the design phase of an accelerator project a considerable amount of calculation and discussion centres around the choice of the transverse focusing system. The pattern of bending and focusing magnets, called the lattice, has a strong influence on the aperture of these magnets which are usually the most expensive single system in the accelerator and which, in turn, can have an important effect on the design of almost all other systems in the synchrotron. We have seen in Fig. 2.2 the pattern of one cell of a simple synchrotron lattice which is repeated many times around the circumference. Passing through this pattern of F and D lenses, particles make betatron oscillations within the envelopes described by β_{h} and β_{v} or, more precisely, the square roots of these quantities.

3.2 The equation of motion

In the last chapter we derived an expression for the change in divergence of a particle passing through the quadrupole. The strength of the quadrupole is characterized by its gradient dB_z/dx, normalized with respect to magnetic rigidity:

$$k = \frac{1}{B\rho}\frac{dB_z}{dx}.$$

If k is negative, the quadrupole is horizontally focusing and vertically defocusing. We first look at the vertical plane. Therefore, the angular deflection given to a particle passing through a short quadrupole of length ds and strength k at a displacement z is

$$dz' = -kz\,ds.$$

We can deduce from this a differential equation for the motion

$$z'' + k(s)z = 0.$$

This is Hill's equation, a second-order linear equation with a periodic coefficient $k(s)$, which describes the distribution of focusing strength around the ring. The above form of Hill's equation applies to motion in the vertical plane; in the horizontal plane,

$$x'' + \left[\frac{1}{\rho(s)^2} - k(s)\right]x = 0.$$

Here the sign before $k(s)$ is reversed interchanging focusing and defocusing. We include an extra focusing term due to the curvature of the orbit which can be significant in small rings and which is the only form of focusing in a constant-gradient synchrotron.

3.3 Solution of Hill's equation

Hill's equation is reminiscent of simple harmonic motion but has a restoring constant $k(s)$ which varies around the accelerator. In order to arrive at a solution, we first assume that $k(s)$ is periodic on the scale of one turn of the ring. The period can also be a smaller unit, the cell, from which the ring is built. The solution, like the differential equation itself, is reminiscent of simple harmonic motion:

$$x = \sqrt{\beta(s)\varepsilon}\cos[\phi(s) + \phi_0].$$

In simple harmonic motion the amplitude is a constant but we see that in addition to $\sqrt{\varepsilon}$, which can be considered an arbitrary constant, there is another amplitude component, a function $\sqrt{\beta(s)}$. Another difference with harmonic motion is that the phase $\phi(s)$ does not advance linearly with time and with distance s around the ring but is a seemingly arbitrary function. Both these functions of s must have the same periodicity as the lattice and they are linked by the condition

$$\phi' = \frac{1}{\beta} \quad \text{or} \quad \phi = \int \frac{ds}{\beta}.$$

We shall later show that this condition is necessary if Hill's equation is to be satisfied, but for the moment let us just accept it. By simple differentiation we can then find

$$x = \sqrt{\beta(s)\varepsilon}\,\cos[\phi(s) + \phi_0],$$

$$x' = -\sqrt{\frac{\varepsilon}{\beta(s)}}\,\sin[\phi(s) + \phi_0] + \left[\frac{\beta'(s)}{2}\right]\sqrt{\frac{\varepsilon}{\beta(s)}}\,\cos[\phi(s) + \phi_0].$$

If we look at this function where $\beta'(s)$ is zero and hence where the second term in the divergence equation is zero, we find an ellipse with semi-axis $\sqrt{\beta\varepsilon}$ in the x-direction and $\sqrt{\varepsilon/\beta(s)}$ in the x'-direction (Fig. 2.7). Its area is $\pi\varepsilon$, where ε is an invariant of the motion for a single particle or the emittance of a beam of many particles.

In older accelerators, constant-gradient machines like the Cosmotron and Bevatron, simple harmonic motion is a very close approximation to reality. In the vertical plane the particles obey the differential equation

$$\frac{d^2 z}{ds^2} + kz = 0,$$

where k, the restoring force per unit displacement, is just related to the gradient of the field, which is constant around the circumference (apart from a few gaps between magnets). By analogy with the wave equation

$$\frac{d^2 z}{ds^2} + \left(\frac{2\pi}{\lambda}\right)^2 z = 0,$$

we can write

$$\left(\frac{2\pi}{\lambda}\right)^2 = k = \frac{1}{B\rho}\frac{\partial B_z}{\partial x}.$$

The solution of such an equation is a wave whose length is λ, namely:

$$z = z_0 \sin\left(\frac{2\pi}{\lambda}\right)s = z_0 \sin\phi.$$

We can see that the derivative of phase is

$$\phi' = \frac{2\pi}{\lambda},$$

but we have already defined $1/\beta$ to be equal to this derivative and can therefore argue that β is a local wavelength of the oscillation. This may help us to understand the way in which β and ϕ vary in the cells of a FODO lattice.

3.3.1 Q value

In order to explore the physical significance of these quantities further, let us look at the definition of the betatron wavenumber Q. Suppose, we again consider a constant-gradient machine. The particle with the largest amplitude in the beam, $\sqrt{\beta\varepsilon}$, starts off with phase ϕ_0, and after one turn its phase has increased by

$$\Delta\phi = \oint \frac{ds}{\beta} = \frac{2\pi R}{\beta}.$$

It has been round the ellipse $\Delta\phi/2\pi$ times. We define the number of such betatron oscillations per turn to be Q. It is also the betatron wavenumber. Using the above relation we see that, for a constant-gradient machine,

$$Q = \frac{\Delta\phi}{2\pi} = \frac{R}{\beta}$$

or

$$\beta = \frac{R}{Q}.$$

This is approximately true for alternating-gradient machines as well, and is often used in juggling machine parameters at the design stage, since Q determines β and hence the beam size.

We shall see later that it is very important that Q is not an integer or a simple fraction, otherwise, over one or more paths around the ellipse, the particle will repeat its path in the machine and see the same field imperfections. These will then build up into a resonant growth. The condition to be avoided is $nQ = p$ (where n and p are integers). This can be done by tuning the restoring gradients of the quadrupoles.

3.4 Matrix description

From now on we deal only with alternating-gradient machines in which the ring is a repetitive pattern of focusing fields, the lattice. Each lattice element may be expressed by a matrix.

Whole sections of the ring which transport the beam from place to place may also be represented as a matrix. Any linear differential equation, like Hill's equation, has solutions which can be traced from one point, s_1, to another, s_2, by a 2×2 matrix, the transport matrix:

$$\begin{pmatrix} y(s_2) \\ y'(s_2) \end{pmatrix} = \begin{pmatrix} a & b \\ c & d \end{pmatrix} \begin{pmatrix} y(s_1) \\ y'(s_1) \end{pmatrix} = M_{21} \begin{pmatrix} y(s_1) \\ y'(s_1) \end{pmatrix}.$$

We shall see later that M_{21} has a rather simple form for each focusing quadrupole that the particle encounters and for the drift length between quadrupoles and it is easy to compute the four elements numerically once we define the length and focusing strength. We can trace particles by simply forming the product of these elementary matrices. But there is also a general relation between the elements a, b, c, and d and the amplitude and phase of transverse motion between any two points. Each term in M_{21} must be a particular function of $\beta(s)$ and $\phi(s)$. The functions $\beta(s)$ and $\phi(s)$ may be calculated by comparing the numerical result of multiplying the individual matrices for quadrupoles and drift lengths with what we know must be the general form of each element. Our first job is to derive the general form of a periodic transport matrix.

We shall simplify the notation by dropping the explicit dependence of β and ϕ on s from the expressions—we will just have to remember that they vary with s. We also introduce a new quantity:

$$w = \sqrt{\beta}.$$

In this new notation we can write the solution of the Hill equation:

$$y = \varepsilon^{1/2} w \cos(\phi + \phi_0).$$

By taking the derivative and substituting $\phi' = 1/\beta = 1/w^2$, we have

$$y' = \varepsilon^{1/2} w' \cos(\phi + \phi_0) - \frac{\varepsilon^{1/2}}{w} \sin(\phi + \phi_0).$$

The next step is to substitute these explicit expressions for y and y' in both sides of the matrix equation. We do this first with the initial condition $\varphi_0 = 0$ (this is the so-called 'cosine' solution) and then again for the 'sine' solution with $\varphi_0 = \pi/2$. This is exactly equivalent to tracing the paraxial and central rays through an optical lens. We write $\phi_2 - \phi_1 = \phi$ for each case. Each of the two conditions gives us two equations for y and y' and thus we obtain four simultaneous equations which can be solved for a, b, c, and d in terms of w, w', and φ. The result is the most general form of the transport matrix:

$$M_{12} = \begin{pmatrix} \dfrac{w_2}{w_1}\cos\phi - w_2 w_1' \sin\phi & w_1 w_2 \sin\phi \\ -\dfrac{1 + w_1 w_1' w_2 w_2'}{w_1 w_2}\sin\phi - \left(\dfrac{w_1'}{w_2} - \dfrac{w_2'}{w_1}\right)\cos\phi & \dfrac{w_1}{w_2}\cos\phi + w_1 w_2' \sin\phi \end{pmatrix}.$$

At first glance, this seems to have complicated the issue but we still have some constraints to apply. The first of these is to restrict M to be between two identical points in successive turns or cells of a periodic structure. This forces $w_2 = w_1$, $w_2' = w_1'$, and φ to become μ, the phase advance per cell. Then,

$$M = \begin{pmatrix} \cos\mu - ww'\sin\mu & w^2 \sin\mu \\ -\dfrac{1 + w^2 w'^2}{w^2}\sin\mu & \cos\mu + ww'\sin\mu \end{pmatrix}.$$

The next simplification is to invent some new functions of β or

$$\alpha = -ww' = -\frac{\beta'}{2},$$
$$\beta = w^2,$$
$$\gamma = \frac{1 + (ww')^2}{w^2} = \frac{1 + \alpha^2}{\beta}.$$

These functions (which are nothing to do with special relativity!) are a complete and compact description of the dynamics. The matrix now becomes even simpler:

$$M = \begin{pmatrix} \cos\mu + \alpha\sin\mu & \beta\sin\mu \\ -\gamma\sin\mu & \cos\mu - \alpha\sin\mu \end{pmatrix} = \begin{pmatrix} a & b \\ c & d \end{pmatrix}.$$

This is the Twiss matrix. It is the basic matrix for periodic lattices and should be memorized.

3.4.1 Stability

Earlier, when describing the physical picture of alternating-gradient focusing, we likened it to a gutter or the brim of a hat. Focusing and defocusing quadrupoles correspond to concave and convex curvature of the brim, respectively. Particles would only remain focused if they were close to the axis when passing through defocusing quadrupoles. We are now in a position to define this condition quantitatively. In the matrix terms, stability is only assured if the product,

$$\{M(s)\}^{Nk}$$

does not diverge after the N periods which make one turn, and the k turns which define an essentially infinite stable life.

Let us write $\boldsymbol{Y}$ as the vector (y, y'). Then the eigenvalues of the matrix $\boldsymbol{M}(s)$ are numbers for which

$$\boldsymbol{MY} = \lambda \boldsymbol{Y}.$$

The eigenvalues are obtained by solving the determinant equation

$$\det(\boldsymbol{M} - \lambda I) = 0,$$

which gives

$$\lambda^2 - \lambda(a + d) + 1 = 0.$$

Here we have used the fact that $\det \boldsymbol{M} = 1$. Writing

$$\cos\mu = \tfrac{1}{2}\mathrm{Tr}\,\boldsymbol{M} = \tfrac{1}{2}(a + d),$$

we find

$$\lambda = \cos\mu \pm i\sin\mu = e^{\pm i\mu}.$$

For stability μ must be real and this implies both

$$\left|\tfrac{1}{2}\mathrm{Tr}\,\boldsymbol{M}\right| \leq 1$$

and

$$|\lambda| = 1,$$

where λ is complex. This test may be applied numerically on any computed value of M for a period to probe stability.

It has been necessary to digress a little to treat this important concept of stability but we now return to the alternative view of M as a sum of many elemental matrices.

3.4.2 The Twiss matrix

Returning now to the Twiss matrix

$$M = \begin{pmatrix} \cos\mu + \alpha\sin\mu & \beta\sin\mu \\ -\gamma\sin\mu & \cos\mu - \alpha\sin\mu \end{pmatrix} = \begin{pmatrix} a & b \\ c & d \end{pmatrix},$$

if we can only find an independent way of computing the numerical values of a, b, c, and d then we may compute μ, β, α, and γ

$$\cos\mu = \frac{\operatorname{Tr}\boldsymbol{M}}{2} = \frac{a+d}{2},$$

$$\beta = \frac{b}{\sin\mu} > 0,$$

$$\alpha = \frac{a-d}{2\sin\mu},$$

$$\gamma = -\frac{c}{\sin\mu}.$$

The values of μ, β, α, and γ are local and apply to the point chosen in the period as a starting and finishing point. We shall see that each individual component, quadrupole, dipole, or drift space in the ring, has its own matrix and this provides the independent method of calculation. We must first choose the starting point, the location, s, where we wish to know β and the other Twiss parameters. By starting at that point in the ring and multiplying the element matrices together for one turn, we are able to find a, b, c, and d numerically for that point. We can then apply the above four equations to find the Twiss matrix. If the machine has a natural symmetry in which there are a number of identical periods, it is sufficient to do the multiplication up to the corresponding point in the next period. The values of α, β, and γ would be the same if we went on for the whole ring. By choosing different starting points we can trace $\beta(s)$ and $\alpha(s)$. We now give the matrices for these three different lattice elements.

3.4.3 Transport matrices for the components of a period

The simplest of these component matrices is that for an empty space or drift length. Figure 3.1(a) shows the analogy between a particle trajectory and a diverging ray in optics. The angle of the ray and the divergence of the trajectory are related:

$$\theta = \tan^{-1}(x').$$

The effect of a drift length in phase space is a simple horizontal translation from (x, x') to $(x + lx', x')$ and can therefore be written as a matrix:

$$\begin{pmatrix} x_2 \\ x_2' \end{pmatrix} = \begin{pmatrix} 1 & l \\ 0 & 1 \end{pmatrix} \begin{pmatrix} x_1 \\ x_1' \end{pmatrix}.$$

The next simplest case is that of a thin quadrupole magnet of infinitely small length but finite integrated gradient:

$$lk = \frac{l}{B\rho} \cdot \frac{\partial B_z}{\partial x}.$$

Figure 3.1(b) illustrates the optical analogy of a thin quadrupole with a converging lens. A ray, diverging from the focal point, arrives at the lens at a

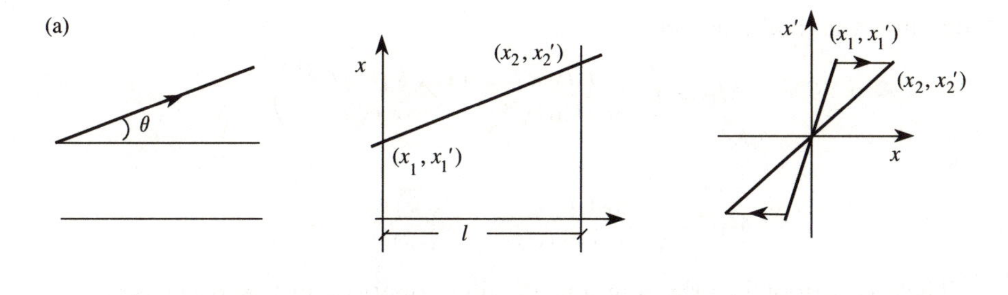

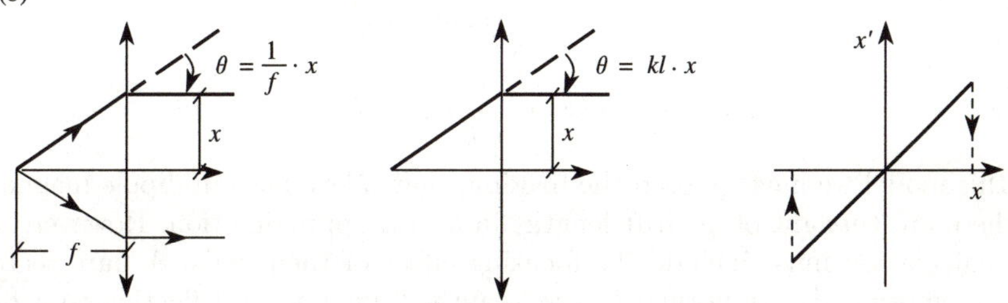

Fig. 3.1 The effect of (a) a drift length and (b) a thin quadrupole seen in real space as an optical ray and a particle trajectory, plotted in phase space and expressed as a transport matrix.

displacement x, and is turned parallel by a deflection

$$\theta = \frac{1}{f} \cdot x.$$

In fact, this deflection will be the same for any ray at displacement x irrespective of its divergence. This behaviour can be expressed by a simple matrix, the thin-lens matrix:

$$\begin{pmatrix} x_2 \\ x_2' \end{pmatrix} = \begin{pmatrix} 1 & 0 \\ -1/f & 1 \end{pmatrix} \begin{pmatrix} x_1 \\ x_1' \end{pmatrix}.$$

A quadrupole has a similar property. A particle arriving at a displacement x obeys Hill's equation:

$$x'' + kx = 0.$$

Hence the small deflection θ is just

$$\Delta x' = -kxl.$$

We see that $lk = 1/f$ is the power of the lens and that the matrix, for a thin lens, can be written as

$$\begin{pmatrix} 1 & 0 \\ -kl & 1 \end{pmatrix}.$$

The lenses of a synchrotron are not normally short compared to their focal length. One must therefore use the matrices for a long quadrupole when one computes the final machine:

$$M_{\mathrm{F}} = \begin{pmatrix} \cos l\sqrt{k} & (1/\sqrt{k})\sin l\sqrt{k} \\ -\sqrt{k}\sin l\sqrt{k} & \cos l\sqrt{k} \end{pmatrix}$$

and

$$M_{\mathrm{D}} = \begin{pmatrix} \cosh l\sqrt{k} & (1/\sqrt{k})\sinh l\sqrt{k} \\ \sqrt{k}\sinh l\sqrt{k} & \cosh l\sqrt{k} \end{pmatrix}.$$

These correspond to the solutions of Hill's equations in F and D cases:

$$z = \cos\sqrt{k}lz_0 + \frac{1}{\sqrt{k}}\sin\sqrt{k}lz_0',$$

$$x = \cosh\sqrt{k}lx_0 + \frac{1}{\sqrt{k}}\sinh\sqrt{k}lx_0'.$$

In this model we have ignored the bending that takes place in dipole magnets and these are thought of as drift lengths in a first approximation. However, an exact calculation must include the focusing effect of their ends. A pure sector magnet, whose ends are normal to the beam will give more deflection to a ray which passes at a displacement x away from the centre of curvature (Fig. 3.2). This particle will have a longer trajectory in the magnet. The effect is exactly like a lens which focuses horizontally but not vertically. The matrices for a sector magnet are

$$M_{\mathrm{h}} = \begin{pmatrix} \cos\theta & \rho\sin\theta \\ -(1/\rho)\sin\theta & \cos\theta \end{pmatrix},$$

$$M_{\mathrm{v}} = \begin{pmatrix} 1 & \rho\theta \\ 0 & 1 \end{pmatrix}.$$

Most bending magnets are not sector magnets but have end faces which are parallel. It is easier to stack laminations this way than on a curve. The entry and exit angles are therefore $\theta/2$ and the horizontal focusing effect is reduced, but

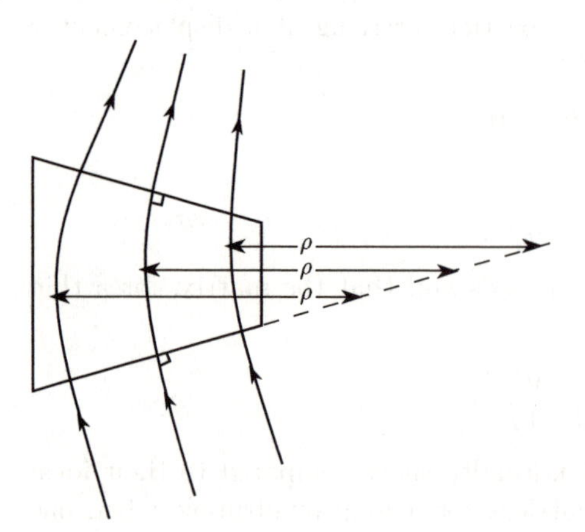

Fig. 3.2 The focusing effect of trajectory length in a pure sector dipole magnet.

there is an additional focusing effect for a particle whose trajectory is displaced vertically. In the computer model, one may convert a pure sector magnet into a parallel-faced magnet by simply adding two thin lenses at each face. They are horizontally defocusing and vertically focusing and their strength is

$$kl = \frac{\tan(\theta/2)}{\rho}.$$

There are further effects from the azimuthal shape of end fields which can be included analytically.

Fortunately, we have computers to help when we come to multiply these elements together to form the matrix for a ring or a period of the lattice (Servranckx and Brown 1984; Garren *et al.* 1985; Schachinger and Talman 1985). A lattice program such as MAD (Iselin and Grote 1991) does all the matrix multiplication to obtain $a, b, c,$ and d from each specified point s, and back again. It prints out β and φ and other lattice variables in each plane, and we can plot the result to find the beam envelope around the machine. This is the way machines are designed. Lengths, gradients, and numbers of FODO normal periods are varied to match the desired beam sizes and Q values.

3.5 Regular FODO lattice

The most convenient way of calculating the numbers a, b, c, and d, which yield the Twiss functions, is to use one of the many computer codes written for the purpose. However, any reader who aspires to become a lattice designer should attempt at least once to solve such a problem analytically. The illustration we shall use for this exercise is a simple FODO lattice (Fig. 3.3) consisting of thin lenses with alternating focal length $1/(kl)$ and spaced by a distance L.

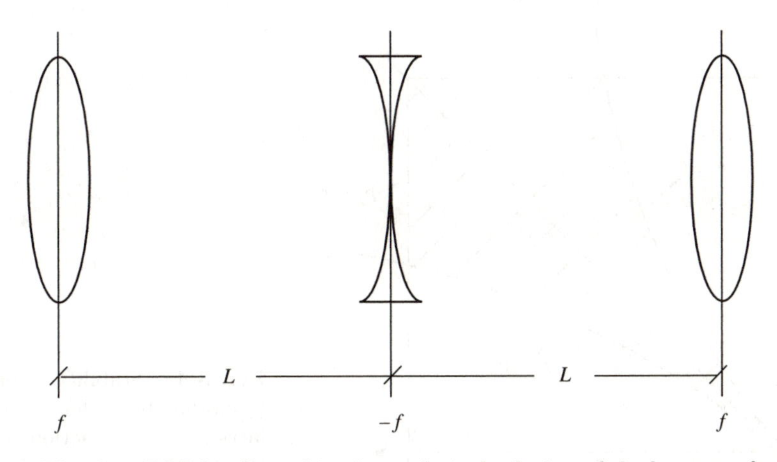

Fig. 3.3 Thin lens FODO cell used in the analytical solution of the betatron function.

The matrix for one period between mid-planes of F lenses is

$$
\begin{aligned}
M &= \begin{pmatrix} 1 & 0 \\ \mp 1/2f & 1 \end{pmatrix} \begin{pmatrix} 1 & L \\ 0 & 1 \end{pmatrix} \begin{pmatrix} 1 & 0 \\ \pm 1/f & 1 \end{pmatrix} \begin{pmatrix} 1 & L \\ 0 & 1 \end{pmatrix} \begin{pmatrix} 1 & 0 \\ \mp 1/2f & 1 \end{pmatrix} \\
&= \begin{pmatrix} 1 - L^2/2f^2 & 2L(1 \pm L/2f) \\ -l/2f^2(1 \mp L/2f) & 1 - L^2/2f^2 \end{pmatrix} \\
&= \begin{pmatrix} \cos\mu + \alpha \sin\mu & \beta \sin\mu \\ -\gamma \sin\mu & \cos\mu - \alpha \sin\mu \end{pmatrix}.
\end{aligned}
$$

We multiply out the product and equate the terms to the Twiss matrix derived earlier. The upper sign is used for the matrix between mid-planes of F lenses and the lower for mid-D. We then obtain,

$$
\begin{aligned}
\cos\mu &= 1 - \frac{L^2}{2f^2}, \\
\sin\left(\frac{\mu}{2}\right) &= \frac{L}{2f}, \\
\beta &= 2L \frac{[1 \pm \sin(\mu/2)]}{\sin\mu}, \\
\alpha_{x,z} &= 0.
\end{aligned}
$$

We see that α, which is proportional to the slope of the beta function, is zero at the planes of symmetry. Also, since the FODO pattern in the h plane becomes DOFO in the v plane, the two values of β can be thought of, as the maximum

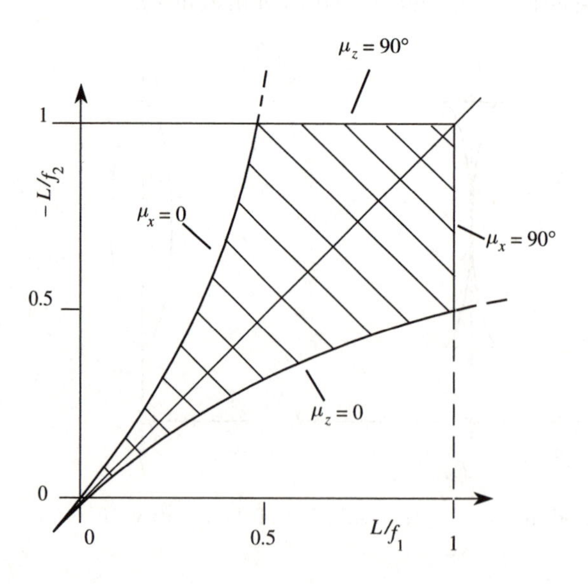

Fig. 3.4 Stability diagram (cross-hatched) for a FODO lattice as a function of the betatron functions. μ is phase advance per cell.

or minimum values in one plane,

$$\frac{\hat{\beta}}{\breve{\beta}} = \frac{1 + \sin(\mu/2)}{1 - \sin(\mu/2)},$$

or as the values of β_x and β_z at an F quadrupole. By repeating the multiplication from the mid-point between two quadrupoles, one may show that

$$\beta_{\text{mid}} = \frac{L}{\sin\mu}\left(2 - \sin^2\left(\frac{\mu}{2}\right)\right).$$

Such analytic expressions are not only useful in a preliminary survey of parameters but can be used to reveal the range of strengths for which the lattice is stable. If we solve the more general matrix product for a horizontally focusing quadrupole of focal length f_1, and a defocusing quadrupole of focal length f_2, we can plot a stable range of f_1 and f_2 for which $\sin\mu < 1$. This is shown shaded in Fig. 3.4.

Exercises

3.1 Solve Hill's equation:

$$y'' + K(s)y = 0$$

by substituting:

$$y = a\sqrt{\beta(s)}\cos[\phi(s) + \phi_0] \quad \text{with } \phi' = \frac{1}{\beta(s)},$$

demonstrating that a necessary condition is:

$$\tfrac{1}{2}\beta\beta'' - \tfrac{1}{4}\beta'^2 + K\beta^2 = 1.$$

3.2 A quadrupole doublet consists of two lenses of focal length f_1 and f_2 separated by a drift length of l in m. Assume that the lenses are thin and show, by writing the three matrices for the lenses, that the product matrix is

$$M = \begin{pmatrix} 1 - l/f_1 & l \\ -l/f^* & 1 - l/f_2 \end{pmatrix},$$

where

$$\frac{1}{f^*} = \frac{1}{f_1} + \frac{1}{f_2} - \frac{l}{f_1 f_2}.$$

3.3 A FODO cell may be considered to be one such matrix with $f_1 = +2f$ and $f_2 = -2f$ followed (and multiplied) by another cell with $f_1 = -2f$ and $f_2 = +2f$. Using the result of the last question, write down these two matrices and show that the product matrix for a half-cell from mid-F to mid-D quadrupole is

$$M = \begin{pmatrix} 1 - l^2/2f^2 & 2l(1 + l/2f) \\ -(l/2f^2)(1 - l/2f) & 1 - l^2/2f^2 \end{pmatrix}.$$

3.4 The matrix for a FODO period must have the form

$$M = \begin{pmatrix} \cos\mu + \alpha\sin\mu & \beta\sin\mu \\ -\gamma\sin\mu & \cos\mu - \alpha\sin\mu \end{pmatrix}.$$

You are given the following data:

$B\rho = 26.68$,
Quadrupole length $= 0.509\,\mathrm{m}$,
Quadrupole gradient $= 12\,\mathrm{T/m}$,
Distance between quadrupole centres $= 6.545\,\mathrm{m}$.

Take the trace of this matrix and equate it to the result of Question 3.2 to obtain an expression for μ, the phase advance per period. Substitute the data to obtain a numerical value for μ. Now use $M_{12} = \beta\sin\mu$ to find β at the mid-plane of the F quadrupole.

3.5 Without multiplying all the matrices together again, write down the expression for β at a mid-D and evaluate.

3.6 A synchrotron consists of 24 FODO cells with the parameters defined in Exercise 3.4. What will be the Q value? Use the smooth approximation to estimate the mean value of β.

3.7 Use the relation $Q = 2\pi\mu$ to calculate the change in Q for a 1% error in strength of all quadrupoles.

4 Circulating beams

4.1 Liouville's theorem

Particle beams in an accelerator obey a conservation law of phase space known as Liouville's theorem. To understand this law we must think of a beam of particles as a cloud of points within a closed contour in a transverse phase-space diagram (Fig. 4.1). Liouville's theorem tells us that the area within the contour, $A = \int p\,dq$, is conserved. The contour is usually, but not always, an ellipse. In Chapter 2, Fig. 2.7 we came across such an elliptical contour—the locus of a particle's motion at a place where the β function is at a maximum or minimum and where the major and minor axes of the upright ellipse are $\sqrt{\varepsilon\beta}$ and $\sqrt{\varepsilon/\beta}$. We could think of this ellipse as the locus of the particle in the beam which has the maximum amplitude of betatron motion and call its area, $\pi\varepsilon$, the emittance. We usually express emittance in units of π mm mrad. According to Liouville the emittance area will be conserved as the beam circulates in a synchrotron or as it passes down a transport line irrespective of the magnetic focusing or the bending

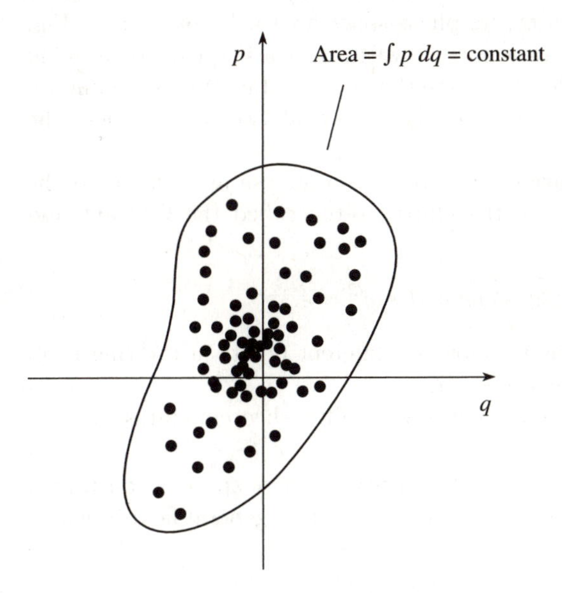

Fig. 4.1 Liouville's theorem applies to this ellipse.

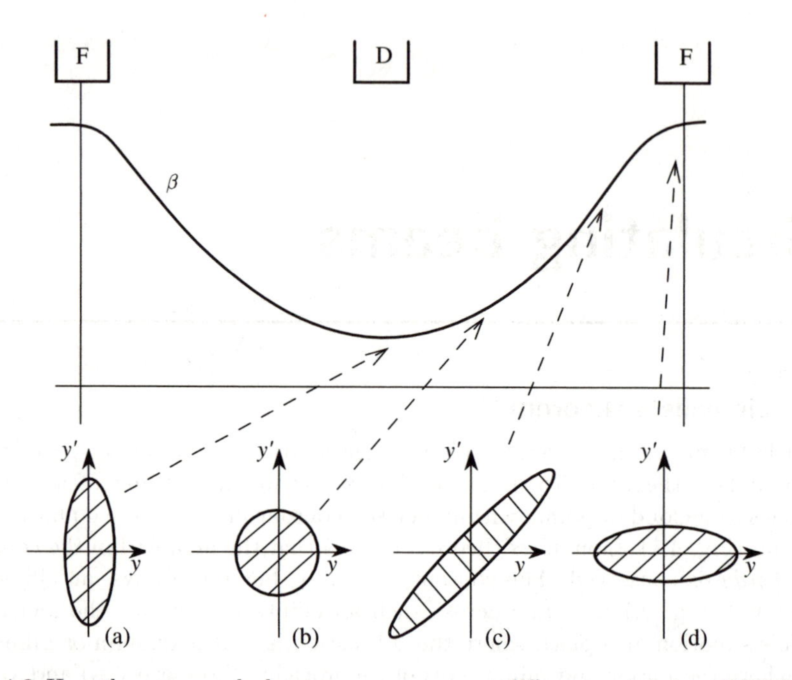

Fig. 4.2 How the conserved phase space appears at different points in a FODO cell. The development of a constant-emittance beam in phase space at (a) a narrow waist, (b) and (c) places where the beam is diverging, and (d) at a broad maximum at the centre of an F lens.

operation performed on the beam. Even though the ellipse may appear to have many shapes around the accelerator, its phase-space area will not change (Fig. 4.2). Its divergence will be large at a narrow waist, near a D quadrupole (a) in Fig. 4.2, while in an F quadrupole (d), where the betatron function is maximum, its divergence will be small. The beam is also shown at two points where the beam is diverging.

In Fig. 4.3 we see how the various features of the ellipse are related to the Twiss parameters. The equation of the ellipse, often called the Courant and Snyder invariant, has the form

$$\gamma(s)y^2 + 2\alpha(s)yy' + \beta(s)y'^2 = \varepsilon.$$

The invariance of this quantity as we move to different points in the ring is an alternative statement of Liouville's theorem.

One word of caution—the strict version of Liouville's theorem states that

> 'In the vicinity of a particle, the particle density in phase space is constant if the particles move in an external magnetic field or in a general field in which the forces do not depend upon velocity.'

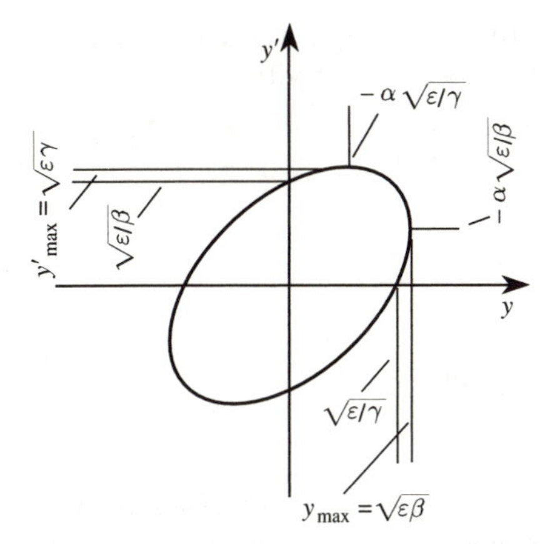

Fig. 4.3 The parameters of a phase-space ellipse containing an emittance ε at a point in the lattice between quadrupoles.

and although this implies conservation of phase-space area, it rules out the application of Liouville's theorem to situations in which space-charge forces within the beam play a role or when particles emit synchrotron light—a velocity-dependent effect.

However, with these precautions in mind we may reliably apply Liouville theorem to proton beams which do not normally emit synchrotron light and to electrons travelling for a few turns in a synchrotron. This is usually too short a time for electrons to emit enough synchrotron light energy to affect their transverse motion.

Now we must ask the question whether Liouville's theorem applies as a proton beam is accelerated. Observations suggest that this is not the case. The beam appears to shrink. However, this is because the coordinates we have used so far, y and y', are not 'canonical' in the sense defined by Hamilton in his mechanics, which is part and parcel of Liouville's mathematical theory of dynamics. We should therefore express emittance in Hamilton's canonical phase space (q, p) and relate this carefully to the coordinates, displacement y, and divergence y', which we have been using so far. We can then define an emittance which is conserved as we accelerate.

In the arguments that follow we shall have to be particularly careful not to confuse Twiss parameters β and γ with the parameters of special relativity, which have the meaning v/c and E/E_0, that is, the velocity divided by that of light, and the total energy divided by the rest energy. To be sure the reader will have to examine the context. For readers who are not familiar with Hamiltonian mechanics, it is sufficient to know that the canonical coordinates of relativistic mechanics are

$$p = \frac{m_0 \dot{y}}{\sqrt{1 - v^2/c^2}}, \qquad q = y.$$

Here q or y is a general transverse coordinate, p its conjugate momentum and we define β and γ when used in the context of special relativity to be

$$\beta = \frac{v}{c},$$

$$\gamma = \frac{1}{\sqrt{1-\beta^2}},$$

$$m_0 = \text{rest mass},$$

$$c = \text{velocity of light},$$

$$p_0 = m_0 c(\beta\gamma).$$

where p_0 is the momentum in the direction of motion of the particle. We may find the relationship between canonical momentum and divergence from the substitution

$$p = m_0 \frac{dq}{dt}\gamma = m_0 \frac{ds}{dt}\frac{dq}{ds}\gamma = m_0 c(\beta\gamma) y'.$$

By simply writing down the Liouville equation in canonical coordinates we can use the above expression to define a conserved quantity and relate it to the area in (y, y') space,

$$\int p\,dq = m_0 c(\beta\gamma) \int y'\,dy = p_0 \int y'\,dy.$$

Thus the conserved quantity is the emittance, ε, of our transverse phase space multiplied by p_0, which is proportional to $\beta\gamma$. Accelerator physicists often call this the invariant or 'normalized' emittance:

$$\varepsilon^* = (\beta\gamma)\varepsilon \ \ (\pi\,\text{mm}\,\text{mrad}).$$

As acceleration proceeds in a synchrotron, the normalized emittance is conserved and the physical emittance within the right-hand side of the equation must fall inversely with momentum if the whole term is to be conserved. Close to the velocity of light this implies that it is inversely proportional to energy:

$$\text{Emittance} = \pi\varepsilon = \int y'\,dy = \frac{\pi\varepsilon^*}{(\beta\gamma)} \propto \frac{1}{p_0}.$$

We therefore expect the beam dimensions to shrink as $1/p_0^{1/2}$ (Fig. 4.4), a phenomenon called 'adiabatic damping'.

4.1.1 Chains of accelerators

Adiabatic damping implies that proton accelerators need their full aperture at injection and it is then that their design is most critical. For this reason

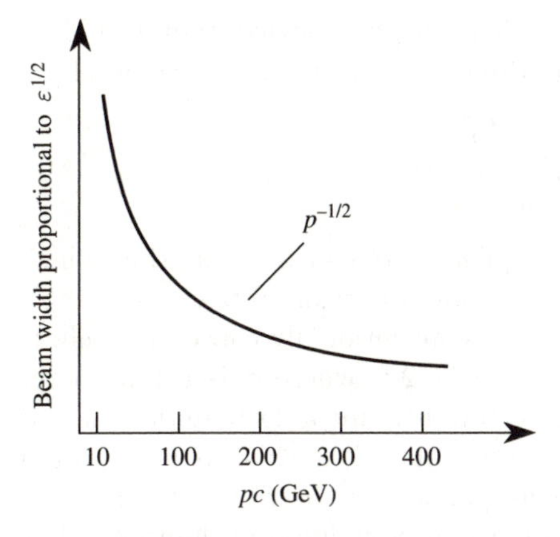

Fig. 4.4 Adiabatic shrinking.

it is economical to split a single large ring into a chain of accelerators—the smaller-radius rings having a large aperture while the higher-energy rings with large radius can have smaller apertures. In these chains of proton accelerators, such as the Linac, PS Booster, PS, the invariant emittance, determined by the parameters of the beam as it leaves the ion source at the beginning of the linac, may be conserved to several hundred GeV. Of course, one must guard against mismatches between machines or non-linear fields which dilate the emittance.

4.1.2 Exceptions to Liouville's theorem

By now, a reader familiar with an electron synchrotron would begin to feel a rising tide of protest. In fact, the invariance of normalized emittance and the shrinking of physical emittance with energy is quite the opposite of what happens in an electron machine. We have already warned that Liouville's theorem applies only to particles guided by external fields and not to electron machines where particles emit some of their own energy. We shall see later in Chapter 8 that electrons, being lighter than protons and hence more relativistic, emit quanta of radiation as they are accelerated. This quantized emission causes particles to jump around in momentum, and momentum changes couple into both planes of transverse phase space. At the same time, there is a steady tendency for particles near the edge of the emittance to lose transverse energy and fall back towards the centre. In an electron machine the emittance is determined not by the Liouville equation but by the equilibrium between these two effects. In fact, it grows with E^2.

4.2 Beam distribution in real space

Suppose we take a number of protons which have the maximum amplitude present in the beam. They follow trajectories at the perimeter of the ellipse but at any instant have a random distribution of initial phases ϕ_0. If we were able to measure y and y' for each and plot them in phase space, they would lie around the ellipse of area $\pi\varepsilon$ and their coordinates would lie in the range of

$$-\sqrt{\beta\varepsilon} \leq y \leq \sqrt{\beta\varepsilon},$$
$$-\sqrt{\varepsilon\gamma} \leq y' \leq \sqrt{\varepsilon\gamma}.$$

However, particles in a beam are usually distributed in a population which appears Gaussian when it is projected on a vertical or horizontal plane (Fig. 4.5). In a proton machine the emittance boundary is conventionally chosen to include about 90% (strictly 87%) of a Gaussian beam at 2σ, where σ is a parameter describing a standard distribution. In an electron machine a 2σ boundary would be too close to the beam and an aperture stop placed at this distance would rather rapidly absorb most of the beam as particles redistribute themselves, moving temporarily into the tails due to quantum emission and damping. The safe physical boundary for electrons depends on the lifetime required but is in the region of 6σ to 10σ. To save an argument about whether one needs 6σ or 10σ, the emittance which is normally quoted for an electron beam corresponds to the σ of the Gaussian projection. We are then free to choose the number of σ's

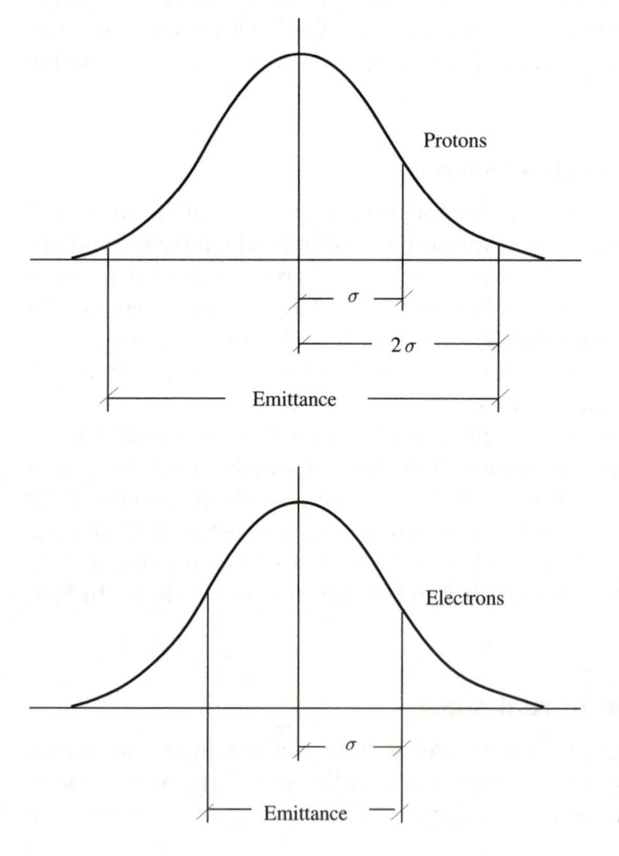

Fig. 4.5 Definitions of emittance for protons (above) and electrons (below).

we must allow. There is consequently a difference between emittance (a factor 4) defined by electron and proton experts. Remembering that particles follow an elliptical trajectory with axes

$$\Delta y = \pm\sqrt{\varepsilon\beta}, \qquad \Delta y' = \pm\sqrt{\varepsilon\gamma}$$

we can write

$$\varepsilon_{\text{protons}} = \frac{(2\sigma)^2}{\beta}$$

and

$$\varepsilon_{\text{electrons}} = \frac{\sigma^2}{\beta}.$$

Beware! Just to complicate matters, some proton specialists have of late taken to using the same, σ^2/β, definition of emittance used for electrons.

The reader is encouraged to consult a number of excellent textbooks for further details or transverse particle dynamics (Steffen 1984; Conte and MacKay 1991; Edwards and Syphers 1993; Bryant and Johnsen 1993).

4.3 Acceptance

In contrast, the acceptance A is unambiguous. It is the size of the hole which is the vacuum chamber transformed into phase space (Fig. 4.6):

$$A = \frac{r^2}{\beta},$$

where r is the semi-axis of the chamber.

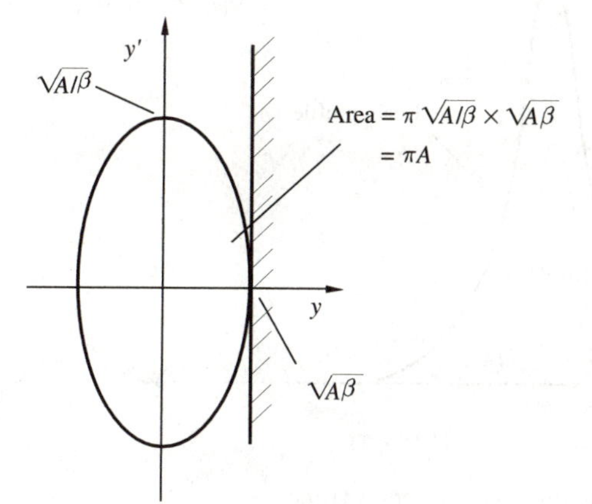

Fig. 4.6 Largest particle grazing an obstacle defines acceptance.

4.4 Measurement of emittance

The methods of observing particle beams are much more limited than the theory might suggest. It is useful to bear this in mind and aim for simplicity in arriving at a design rather than relying upon a complex procedure to tune the machine once it is built. We cannot measure x' and it is not easy to measure x, since a simple capacitive pickup electrode would just give the average beam position $x = 0$. This is because the particles have random phases and their motion is said to be *incoherent.*

In electron machines, observations of beam size are made by refocusing the synchrotron light emitted, though the beam dimensions can be so small that special image-scanning techniques are needed to resolve the image. Of course, this has no effect on the beam. On the other hand, the most reliable and straightforward way to measure proton beam size is to drive a scraper into the beam or move the beam across the scraper and integrate the beam loss curve. Clearly, this is destructive. Another diagnostic tool is a wire which is scanned across the beam very rapidly. Secondary particles generated are counted with a scintillation telescope. Here one must be careful not to dilate the beam by scattering (or burn the wire). One non-destructive instrument for protons, the ionization beam scanner (IBS), produces a profile of the beam as in Fig. 4.7. It shows the displacements of particles in a region about the beam centre and measures $\sqrt{\beta\varepsilon}$, the half-width or envelope of the beam (Fig. 3.1). The principle is shown in Fig. 4.8. A zero electrical potential scans across the beam, allowing electrons to be collected from its surface as the beam ionizes the residual gas.

This IBS can also give a 'mountain range' display of the beam during acceleration (Fig. 4.9) demonstrating adiabatic shrinkage. Its disadvantage is that the

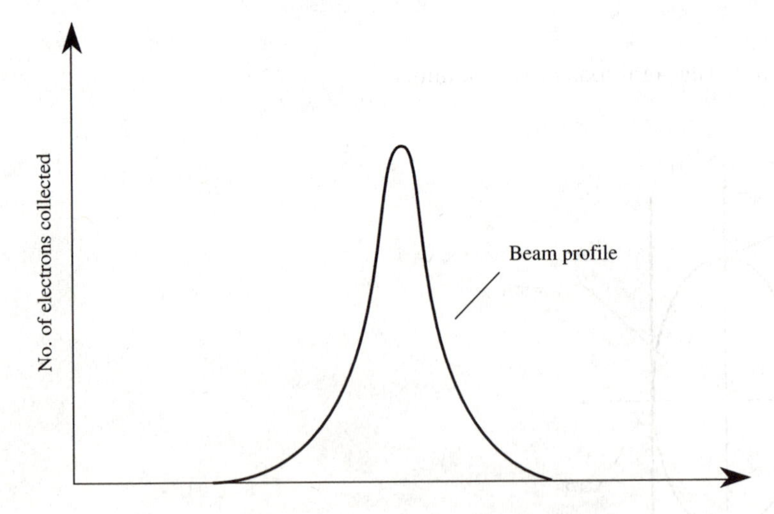

Fig. 4.7 Ideal ion beam scanner scope trace.

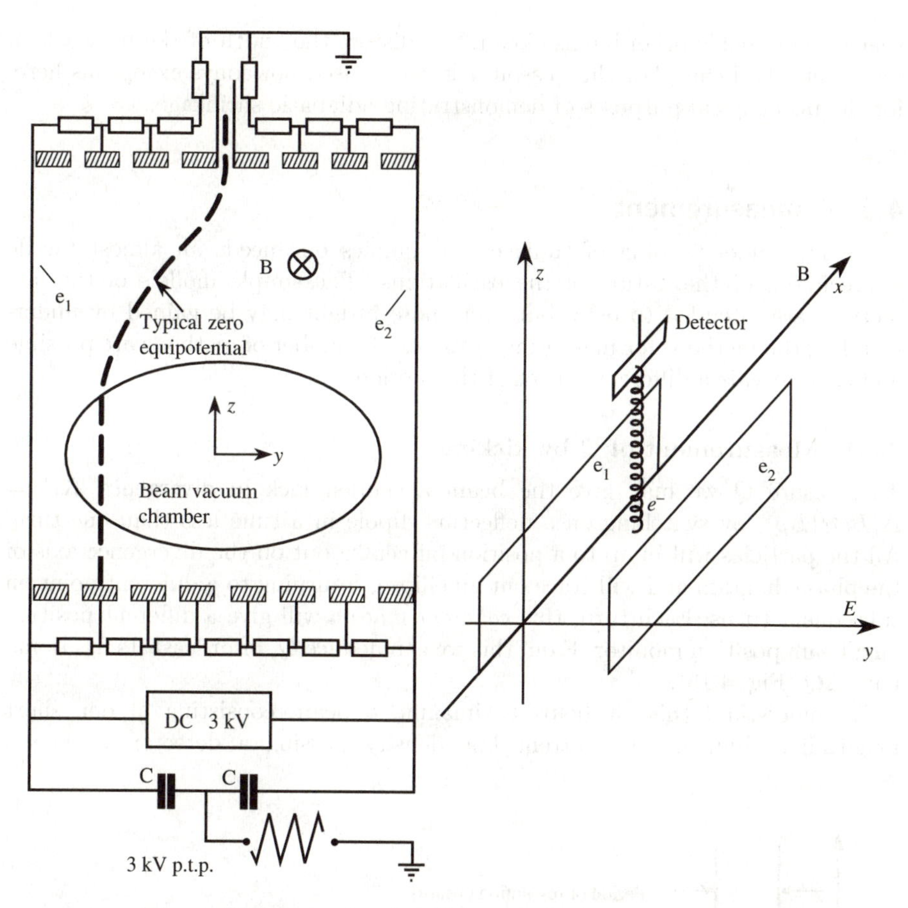

Fig. 4.8 Diagram showing IBS principle of beam scanning.

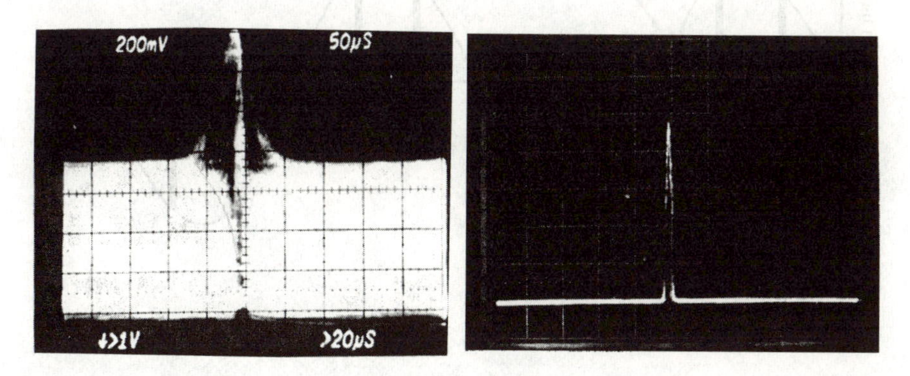

Fig. 4.9 Mountain range display and horizontal profile; 1 cm/division horizontal profile at 350 GeV.

space-charge fields of an intense beam can distort the width of the peak which represents the beam. For this reason it is rarely used nowadays except, as here, for the pedagogical purposes of demonstrating adiabatic shrinkage.

4.5 Q measurement

For a full understanding of transverse dynamics one needs an almost tactile appreciation of the nature of the oscillations. The simple models of the last sections are intended to help, but even more insight may be gained by understanding the methods for measuring Q values. A number of methods are possible and each reveals a different aspect of the motion.

4.5.1 Measurement of Q by kicking

To measure Q we may give the beam a sudden kick in divergence $\Delta x' = \Delta(Bl)/(B\rho)$, by switching on a deflecting dipole in a time less than one turn. All the particles will jump to a position off centre but on the divergence axis of the phase diagram and will trace out an ellipse, returning to a different point on subsequent turns. Each turn, this *coherent motion* will give a different position on a beam position monitor. From this we can deduce Q, or at least its fractional part ΔQ (Fig. 4.10).

To understand this, it helps to imagine a beam consisting of one short longitudinal bunch. The current line density passing a detector is then a

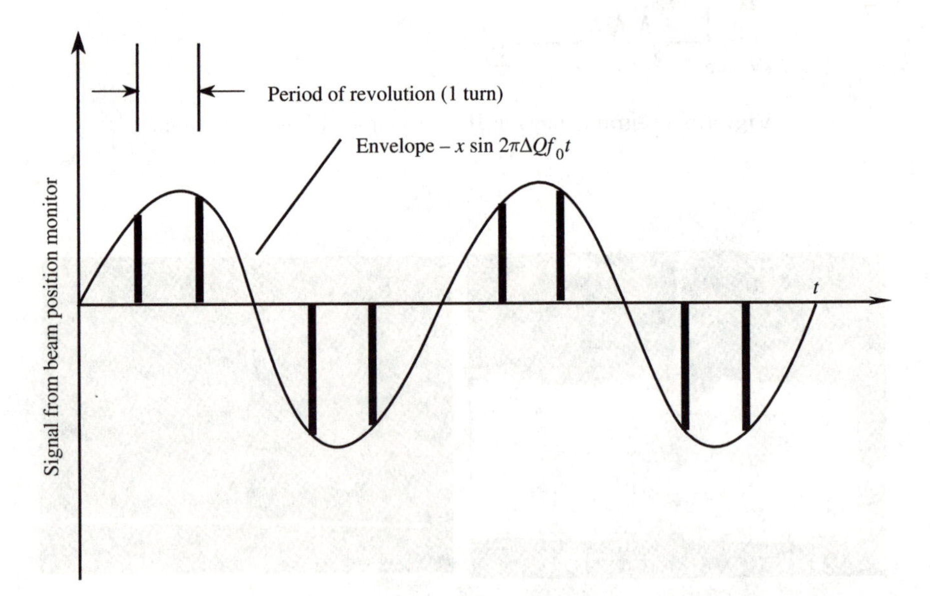

Fig. 4.10 Coherent motion following a sudden kick and observed on a beam position monitor.

Fourier series,

$$\rho(t) = \sum_n a_n \sin 2\pi n f_0 t.$$

The beam position detector sees the betatron oscillations following the kick as

$$y(t) = y_0 \cos 2\pi f_0 Q t$$

but modulated by $\rho(t)$. An oscilloscope connected to the pickup will give a display of the modulated signal, the product of ρ and y. We make use of the elementary relation

$$\sin a \, \cos b = \tfrac{1}{2}[\sin(a+b) + \sin(a-b)]$$

and obtain

$$\rho(t)y(t) = \frac{1}{2}\sum_n a_n y_0[\sin 2\pi(n+Q)f_0 t + \sin 2\pi(n-Q)f_0 t].$$

The envelope of the oscilloscope signal is the slowest of these terms, in which $(n - Q)$ is the fractional part of Q. The other terms in the series reconstruct the spikes in the signal occurring once per turn. The fractional part of Q is the reciprocal of the number of spikes in one wavelength of the envelope.

4.5.2 Knockout and Q measurement

A very simple form of resonance can be induced by applying a deflecting field with the frequency of a betatron sideband, as the frequencies $(n \pm Q)f_0$ are called. We can invert the treatment above to show that if you apply a signal of the form

$$\sin[2\pi(n-Q)f_0 t],$$

then the particle passing the electrodes of the deflector every turn experiences a kick

$$\sin[2\pi n f_0 t] \sin[2\pi(n-Q)f_0 t].$$

Using again the elementary trigonometry of the last section, we see that the $(a - b)$ component from mixing these two frequencies is in resonance with the betatron motion

$$\cos 2\pi Q f_0 t.$$

Once in resonance, the particle is deflected on each turn by a kick, which increases its amplitude of transverse oscillations and is in phase with the excitation so that it blows up the beam. Note that this, like the previous method, gives a value of the fractional part of Q with respect to the nearest integer but gives no information about which integer this is.

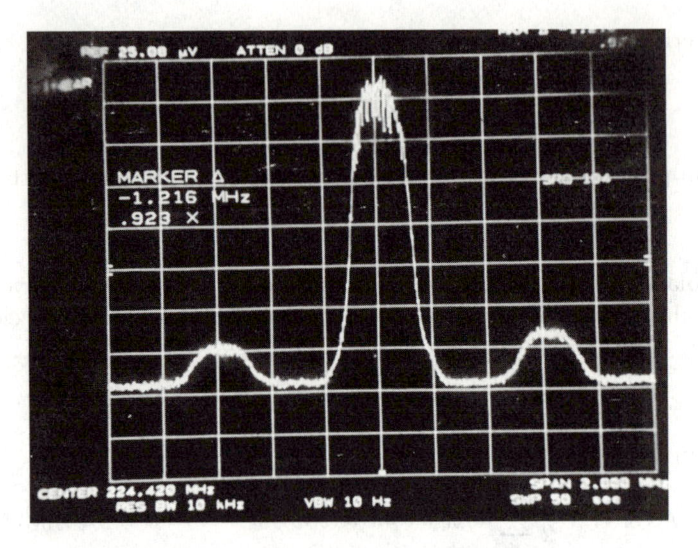

Fig. 4.11 Transverse pickup seen on a spectrum analyser.

4.5.3 Measurement by analysing the frequencies emitted by the beam

It is also possible nowadays to detect betatron frequencies in the statistical noise signal detected by a simple transverse pickup, which can be just a pair of plates. This is then displayed with an integrating spectrum analyser, which is really a scanning radio receiver connected to an oscilloscope. Peaks appear as sidebands to the revolution frequency in the display of response versus frequency (Fig. 4.11). Their separation is $2\Delta Q f_0$ where ΔQ is the fractional part of Q.

Readers may wonder why the particles, evenly spread around the ellipse in phase space, can generate such a signal. The answer is that they are finite in number and the pickup samples but a small fraction of them. In a sample there are always significant statistical fluctuations of the centre of charge, or mean displacement, which the spectrum analyser picks up.

Exercises

4.1 The emittance of a proton beam at injection in the SPS is 2π mm mrad. Calculate the half-width of the beam at an F quadrupole where $\beta = 108$ m.

4.2 What is the maximum value of the divergence in the beam if the β at a defocusing quadrupole is 18 m and $\alpha = 0$?

4.3 What is the normalized emittance of this beam if the above data refer to a proton momentum of 10 GeV/c?

4.4 If this normalized emittance is accelerated to 400 GeV/c, what will be the half-width of the beam at an F quadrupole ($\beta = 109$ m)?

4.5 A synchrotron has a mean radius of 1100 m. Protons circulate at a momentum of 10 GeV/c. What is the revolution frequency?

4.6 What is the separation (in Hz) of the two betatron sidebands nearest to the seventh revolution frequency harmonic ($Q = 6.47$)?

4.7 A beam is disturbed by a transverse kick producing the signal on a transverse pickup shown in Fig. 4.10. What is the fractional part of Q?

5 Longitudinal dynamics

5.1 Longitudinal motion

We shall devote the whole of Chapter 10 to a description of r.f. accelerating cavities but, for the purposes of this chapter, we need only think of the accelerating system as a simple resonant 'pill-box'. We start by defining a few technical terms.

5.1.1 Voltage per turn

This is the total potential difference experienced by the particle due to the r.f. field across all the accelerating structures in the ring. It is just the integrated voltage, or energy gained by the particle, as it passes through the cavities on one turn of the machine.

5.1.2 The r.f. frequency programme and controls

The bending radius, ρ , should remain constant in a synchrotron as acceleration proceeds. To achieve this the particle's momentum must be incremented on each turn by a precise voltage:

$$V = V_0 \sin \phi_{\mathrm{s}}$$

which is just sufficient to keep pace with the rate of rise of the magnetic guide field $\dot{B}$. The amplitude V_0 is pre-programmed and controlled by automatic voltage control in the low-level control circuitry of the transmitter which powers the cavity. The synchronous phase ϕ_{s} is controlled by another servo system which compares the phase of the r.f. voltage with the passage of the bunch. The bunch is detected with a beam monitor which must have a wide band of frequency response to follow the changes in particle revolution frequency. The frequency of the low-level oscillator which provides the input for the power amplifier feeding the cavity must also be pre-programmed and controlled to follow the change in the velocity of the particle. Finally, the radial position of the bunch in the vacuum chamber is measured with a beam position monitor and yet another servo system corrects the transmitter frequency from its programmed value if the beam is found to be displaced from the central orbit.

5.1.3 Harmonic number

A particle circulates around the machine with a period or frequency:

$$\tau = \frac{C}{\beta c}, \qquad f = \frac{\beta c}{2\pi R},$$

where $C = 2\pi R$ is the circumference and βc is the velocity. The *synchronous particle* is defined as that particle which always arrives at the desired synchronous phase lag ϕ_s behind the rising zero-crossing of the r.f. wave. For this to occur, the r.f. frequency, f_a, must be an integer multiple of f:

$$f_a = hf,$$

where h is known as the *harmonic number*. The integer h is often chosen for practical reasons: to make the r.f. frequency high, so that it falls in a band where amplifiers and other components are readily available, or sometimes just to keep the dimensions of the cavities and wave-guides reasonably small.

5.1.4 Bunches and buckets

In big modern synchrotrons, h may be very large. For example, for LEP, $h =$ 31 320 and there were 31 320 places on the circumference where a particle could be located and arrive synchronously. The segments of the circumference centred on these points are called *buckets*. The groups of particles in these buckets are called *bunches*. Not all buckets need to be filled with bunches. For example, when LEP operated as a collider, only four or eight equally spaced buckets were filled in order to maximize the probability of collision.

Imagine the cylindrical coordinate system of Fig. 5.1, rotating with the velocity of a bunch of particles. Some particles arrive after the zero-crossing and therefore lag behind the r.f. wave by a phase angle ϕ; others arrive before. The synchronous particle is timed to arrive at $\phi_s = 0$.

We may plot a particle's motion in 'longitudinal phase space'. One of the two coordinates is the energy difference between the particle and the other, the ideal, or synchronous, particle.

The other, plotted along the horizontal axis (Fig. 5.1) is the relative phase of the particle and r.f. wave. Note that a particle outside the range $-\pi < \phi < \pi$ falls in one of the other h segments of the circumference, and its phase can always be redefined with respect to the nearest rising zero-crossing, to keep ϕ within this range. Particles follow paths in this phase space that are either closed and therefore stable, or open and unstable. Readers who are worried about the use of ϕ to describe a displacement in a phase-space diagram may reflect that a phase advance is just the distance on our rotating cylinder between the proton and the origin. Alternatively, we may think of it as the time of arrival.

5.1.5 Stability of the lagging particle

We suppose that the particle velocities are well below the velocity of light. A particle B which arrives late receives an extra energy increment, which will cause

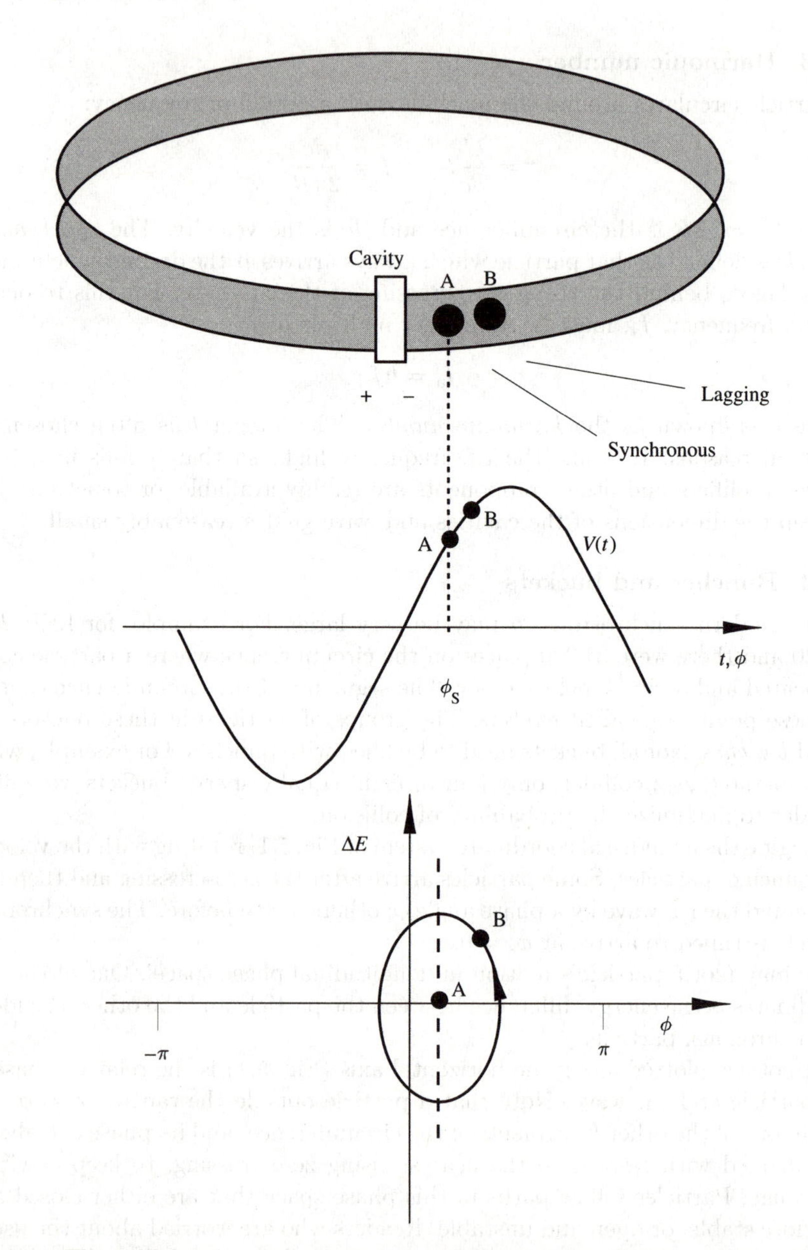

Fig. 5.1 The cylindrical coordinate system which rotates with beam demonstrating the meaning of r.f. phase angle in longitudinal phase space.

it to speed up and overtake the synchronous particle, A. In doing so, its energy defect ΔE, grows and, provided the amplitude is not too large, its trajectory will follow an ellipse in phase space. This describes its motion up and down the r.f. wave (Fig. 5.2) and may remind readers of the representation of a simple

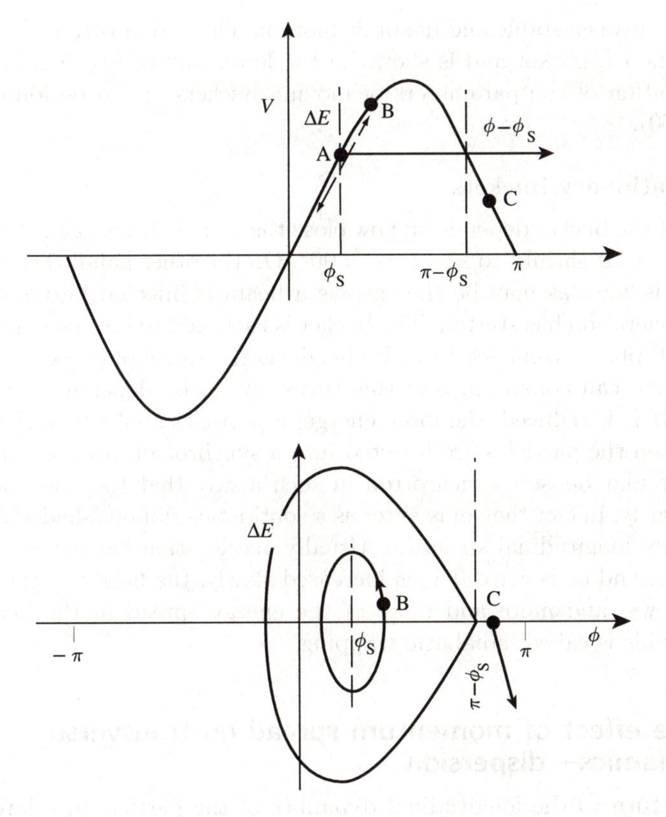

Fig. 5.2 The limiting trajectory for a particle in a 'moving' or accelerating bucket when the stable phase is not zero.

harmonic oscillator, or pendulum, when plotted in a phase-space diagram of velocity versus displacement. The trajectory is closed and over many turns the average deviation from the synchronous energy is zero. This *phase stability* depends upon the fact that δE is positive when $\phi - \phi_{\mathrm{S}}$ is small and positive (Montague 1977; Le Duff 1992).

Even if the oscillation is so large that the particle reaches the non-linear part of the r.f. wave and over the top of the wave, it will still be restored and oscillate about the stable phase provided it does not reach and pass the point where it receives less voltage than the synchronous particle. On this non-linear part of the curve the motion is no longer an ellipse but is distorted into a fish-shape but its trajectory is still closed and stable. However, if a particle C oscillates with such large amplitude that it falls below the synchronous voltage, an increase in ϕ will cause a negative ΔE, which causes ϕ to move further away from the synchrotron particle. This particle is then clearly unstable and will be continuously decelerated. There is a particle which, starting at $\phi = \pi - \phi_{\mathrm{S}}$, would trace out a limiting fish-shaped trajectory, which is the boundary or *separatrix* between stable and unstable motion. The region within this separatrix is called the r.f. bucket and is shown in the lower half of Fig. 5.2. Formulae for the calculation of the parameters of moving buckets are to be found in Bovet *et al.* (1970).

5.1.6 Stationary buckets

The size of the bucket depends on how close the stable phase, ϕ_s, is to the crest of the sine wave. It shrinks to zero if $\phi_s = 90°$. On the other hand, there is a special case if ϕ_s is zero, as may be the case as a beam is injected into a synchrotron before acceleration has started. The bucket is then said to be 'stationary' stretching over all phases from $-\pi$ to π. Its height is the range of energies $2\Delta E$ which the r.f. wave can constrain, and this turns out to be dependent on $\sqrt{V}$ for a given ϕ_s. If V is reduced, the more energetic particles spill out of the bucket.

Very often the particles are injected into a synchrotron from a linear accelerator or circular booster synchrotron in such a way that they do not all arrive synchronously. In fact they may enter as a continuous ribbon (shaded in Fig. 5.3), without any longitudinal structure. Usually acceleration has not yet started, B is constant, and ϕ_s is zero. If V is increased slowly, the height of the stationary bucket grows, and more and more of the energy spread in the beam, ΔE is trapped. This is called 'adiabatic trapping'.

5.2 The effect of momentum spread on transverse dynamics—dispersion

We will return to the longitudinal dynamics of the particle in a later chapter. Meanwhile, we are beginning to become too involved with technical matters and should return to examine another fundamental principle of circular machines which controls the motion of particles in the transverse direction. We, therefore, pass now from a study of the motion of a particle with respect to the r.f. wave to consider how its energy oscillations affect its orbit in transverse phase space.

5.2.1 Closed orbit

The bending field of a synchrotron is matched to some ideal (synchronous) momentum p_0. A particle of this momentum and of zero betatron amplitude will pass down the centre of each quadrupole, be bent by exactly 2π by the bending magnets in one turn of the ring, and remain synchronous with the r.f. frequency. Its path is called the closed orbit of the central (or synchronous) momentum particle. In Fig. 2.9 this ideal orbit was the horizontal axis. We see particles executing betatron oscillations about it but these oscillations do not replicate every turn. In contrast, the synchronous orbit closes on itself so that x and x' remain zero.

5.2.2 Gravitational analogy

Before attempting to explain what happens to a particle whose momentum is slightly different from the ideal value, we digress to understand the influence of

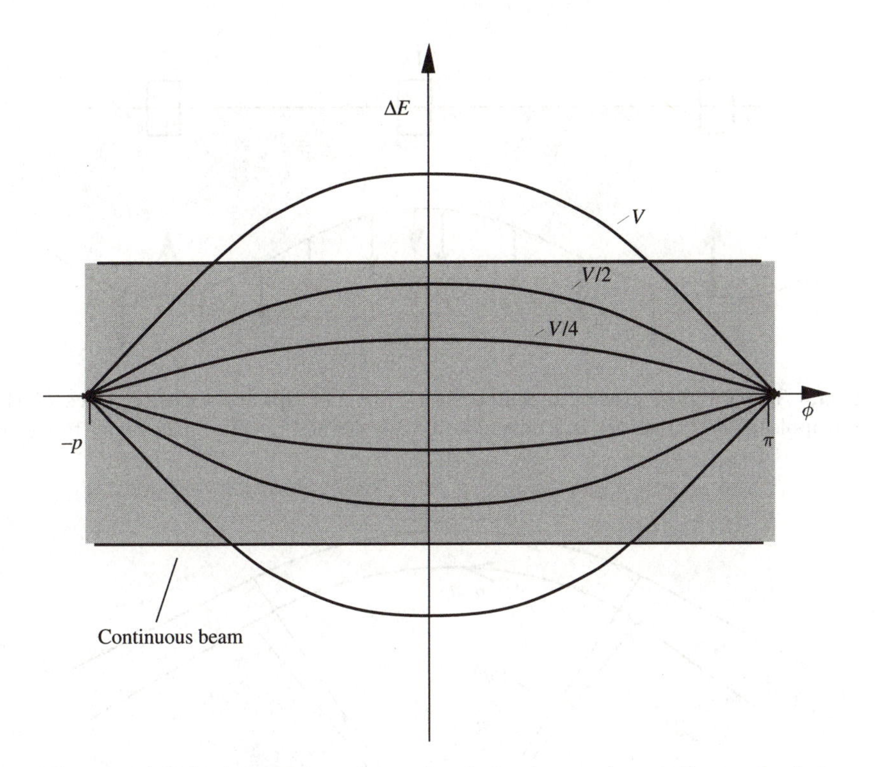

Fig. 5.3 Adiabatic trapping of coasting beam in growing stationary bucket.

another everyday force: gravity. Those new to the field are sometimes fascinated by a question which they dare not ask, 'Why do particles, stored sometimes for days in a synchrotron, not gradually spiral down due to gravity and hit the lower poles of the magnets and how therefore can the central orbit close on itself?'

Let us try to answer this by imagining for a moment that we were able to 'switch on' gravity slowly. The vertical closed orbit of the particle would adjust itself to appear as in Fig. 5.4, an orbit which closes but is everywhere slightly lower than the mid-plane. The vertical deflection from a quadrupole is

$$\Delta z' = \frac{\Delta(lB)}{B\rho} = klz.$$

The continuous force of gravity is then balanced by the vertical focusing forces of the F quadrupoles (less the defocusing forces of the D's).

We have not derived the shape of this orbit but clearly there can be such a closed orbit for which gravity is balanced. This trajectory will be a particular and unique solution of Hill's equation. It is pretty clear from symmetry that the shape of the orbit must be identical in all cells and must be closer to the axis in vertically defocusing lenses. This will ensure that the net focusing is upwards. We might define a 'suspension' function which defines its shape for a particle of unit mass and magnetic rigidity. In practice, the suspended orbit function is only

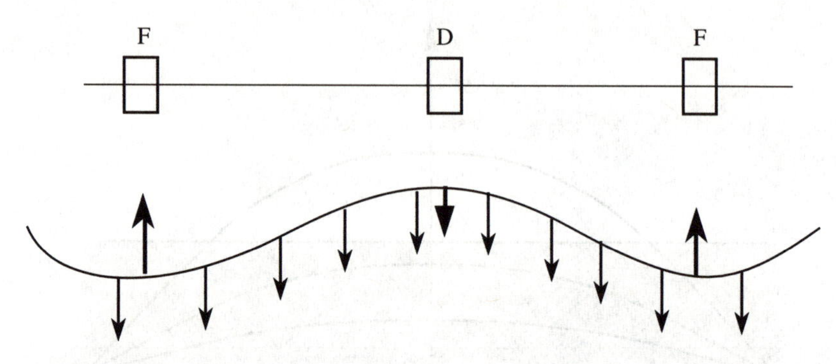

Fig. 5.4 The force of gravity is balanced by the net upward focusing force of the quadrupoles. The trajectory is a new closed orbit defined by the 'suspension' function.

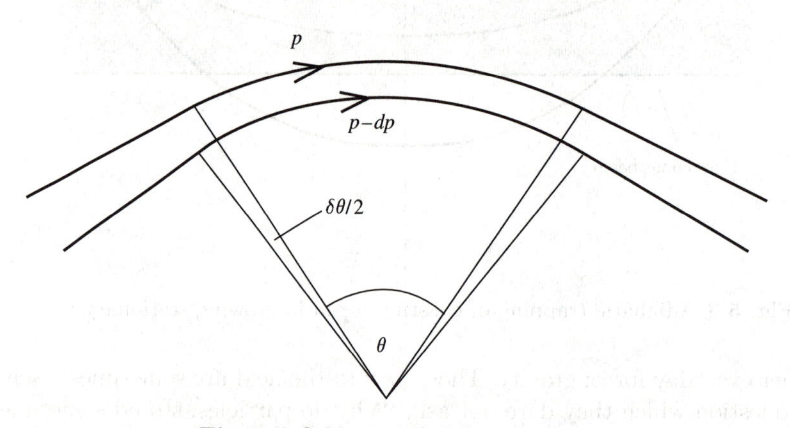

Fig. 5.5 Orbits in a bending magnet.

very slightly different from the mid-plane and the difference may be ignored. Its usefulness here is merely a prelude to the explanation of the quantity 'dispersion' which describes the behaviour of a particle whose momentum defect is $\Delta p/p$ with respect to the synchronous particle.

5.2.3 Orbit of a low-momentum particle

We now take a bird's eye view and look at a closed orbit which is distorted in the horizontal plane due to the particle having a momentum defect .

Figure 5.5 shows a particle with a lower momentum $\Delta p/p < 0$ and which, therefore, is consistently bent horizontally more in each dipole of a FODO lattice. Like the orbit due to gravity, we might argue that the total deflection, being more than 2π, would cause it to spiral in. But again we may argue that there is indeed a closed orbit for this lower momentum in which the extra bending forces are compensated by extra focusing forces as the orbit is displaced inwards in the F quadrupoles and less so in the D's. This is shown in Fig. 5.6. Just as in our

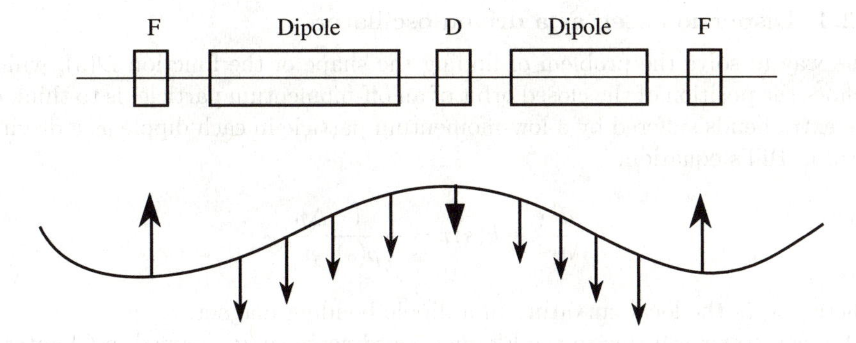

Fig. 5.6 The extra inward force given to a low-momentum particle by the dipoles is balanced by the quadrupoles and defines a dispersion function.

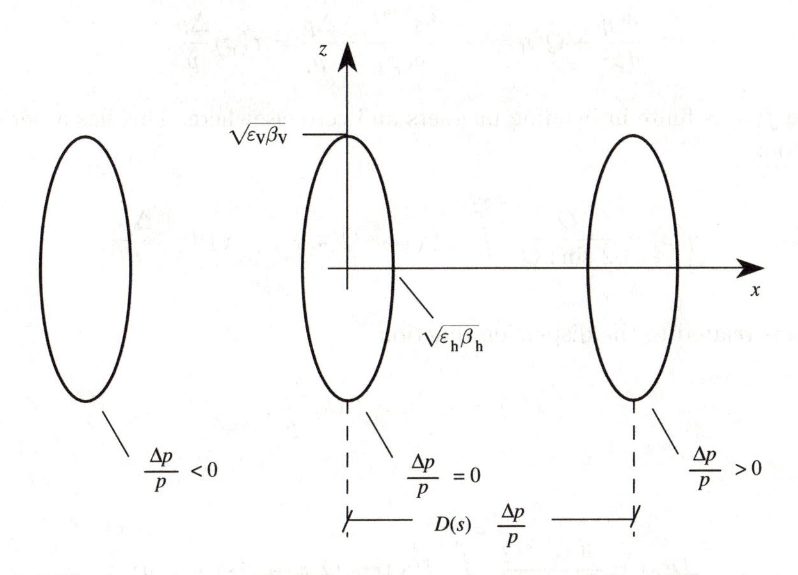

Fig. 5.7 The beam cross sections in real space for beams of three different momenta at a point where the dispersion function is large.

gravitational analogy we may describe the shape of this new closed orbit for a particle of unit $\Delta p/p$ by a *dispersion* function $D(s)$.

This clearly means the beam will be wider if it has momentum spread. In Fig. 5.7 we see how the effect of dispersion for off-momentum orbits combines with betatron motion.

The betatron motion of each of the three kinds of particles $\Delta p/p < 0$, $=0$, and $\Delta p/p > 0$ is confined to an ellipse in physical $(x,\, z)$ space. The ellipses for each momentum are separated by a distance $D(s)\Delta p/p$. Hence, the minimum semi-aperture required for the beam will be

$$a_{\mathrm{v}} = \sqrt{\beta_{\mathrm{v}}\varepsilon_{\mathrm{v}}}, \qquad a_{\mathrm{h}} = \sqrt{\beta_{\mathrm{h}}\varepsilon_{\mathrm{h}}} + \left| D(s)\frac{\Delta p}{p} \right|.$$

5.2.4 Dispersion seen as a driven oscillator

One way to solve the problem of finding the shape of the function $D(s)$, which defines the position of the closed orbit of an off-momentum particle, is to think of the extra bends suffered by a low-momentum particle in each dipole as a driving term in Hill's equation:

$$\frac{d^2x}{ds^2} + k(s)x = -\frac{1}{\rho(s)}\frac{\Delta p}{p},$$

where $\rho(s)$ is the local curvature in a dipole bending magnet.

We can transform this to the Floquet coordinates (η, φ) defined in Chapter 6 and reduce the problem to that of a forced harmonic oscillator,

$$\frac{d^2\eta}{d\varphi^2} + Q^2\eta = -\frac{Q^2\beta^{3/2}(\varphi)}{\rho(\varphi)}\frac{\Delta p}{p} = f(\varphi)\frac{\Delta p}{p},$$

where $f(\phi)$ is finite in bending magnets and zero elsewhere. This has a periodic solution:

$$\eta(\varphi) = \frac{Q}{2\sin\pi Q}\int_{\varphi}^{\varphi+2\pi} f(\chi)\cos Q(\pi + \varphi - \chi)\,d\chi \cdot \frac{\Delta p}{p},$$

which is related to the dispersion function

$$x(s) = \beta^{1/2}\eta = D(s)\frac{\Delta p}{p},$$

with

$$D(s) = \frac{\beta(s)^{1/2}Q}{2\sin\pi Q}\int_{\varphi}^{\varphi+2\pi} f(\chi)\cos Q(\pi + \varphi(s) - \chi)\,d\chi.$$

Alternatively, we may compute $D(s)$ using a 3×3 transport matrix operating on a vector which has $\Delta p/p$ as its third term:

$$\begin{pmatrix} x \\ x' \\ \Delta p/p \end{pmatrix}_2 = \begin{pmatrix} m_{11} & m_{12} & m_{13} \\ m_{21} & m_{22} & m_{23} \\ 0 & 0 & 1 \end{pmatrix}\begin{pmatrix} x \\ x' \\ \Delta p/p \end{pmatrix}_1.$$

The four elements in the top-left corner are just the terms in the familiar 2×2 transport matrix. We can write down a matrix, including the new elements m_{13} and m_{23} for all the magnetic elements of a period just as we did in the case of 2×2 matrices. Choosing a starting point at azimuth s, we can multiply all the matrices together until we have moved exactly one period onward. Just as we were able to deduce the value of β, α, and γ at s from the first four terms m_{11}, m_{12}, m_{21}, and m_{22}, we can now find the slope and value of the dispersion function from

$$D'(s) = \frac{m_{13}m_{21} + (1 - m_{11})m_{23}}{(1 - m_{11})(1 - m_{22}) - m_{21}m_{12}},$$

$$D(s) = \frac{m_{13} + m_{12}D'(s)}{1 - m_{11}}.$$

5.2.5 An example of a dispersion function

In Fig. 5.8 we see the way the dispersion function $D(s)$ varies for one-sixth (18 FODO periods) of the SPS. This 'old-fashioned' machine was designed at a time when it was thought preferable, rather than complicate the pattern with special quadrupoles, to keep a simple regular FODO focusing structure even in the central part of the superperiod, where there had to be room for other machine components. In this region bending magnets were just left out and the pattern of missing bending magnets adjusted to make the right amount of Fourier harmonic to reduce D close to zero.

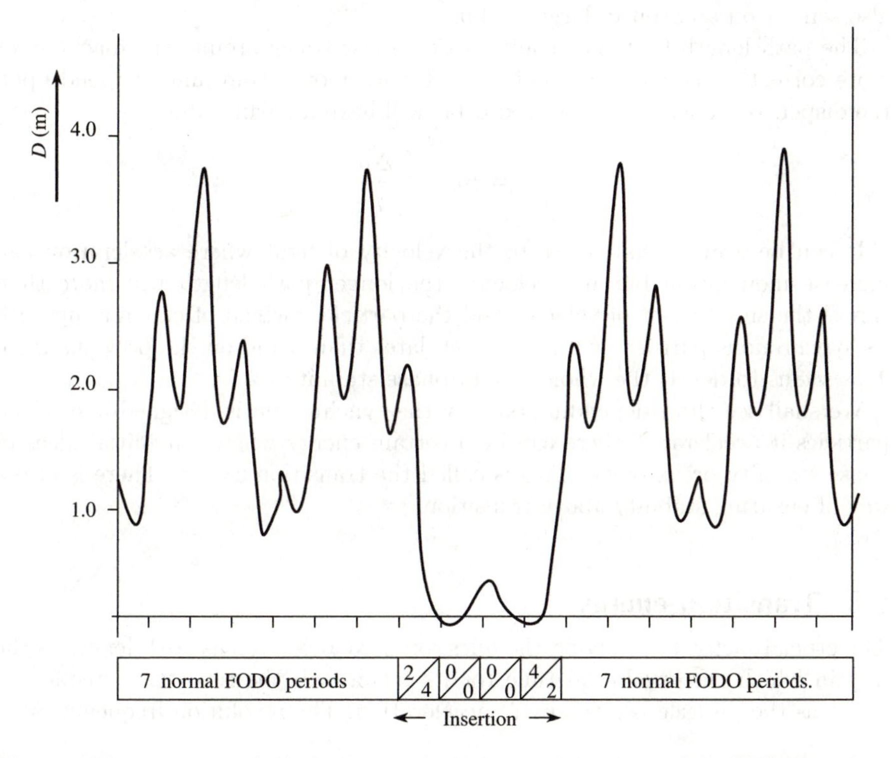

Fig. 5.8 The variation of the dispersion function in one sextant of the SPS centred on the long straight section.

The price one pays is that elsewhere, in the arcs, the Fourier components combine to double the value of D. It illustrates how the Fourier pattern of bending can drive waves in $D(s)$. This was of little importance in a machine of rather small energy spread and large horizontal aperture. In a modern machine, gaps and inserts in the bending pattern would be introduced without exciting the 'beating' of $D(s)$ in the arcs of the machine. A special focusing insertion can be designed which acts rather like a telephoto lens of quadrupoles on either side of the centre of the superperiod. With modern optical matching programs one can make $D(s)$ follow a regular pattern in the arcs where there are dipole magnets and bring it down to zero in the centre of the superperiod, where modern colliders usually require the smallest possible beam size (Brinkmann 1987; Bryant 1992).

5.2.6 Phase stability revisited

When we earlier described the principle of the phase stability of the lagging particle, we used the argument that a particle, arriving late because of its lower energy, would see a higher r.f. voltage from the rising waveform and, accelerated to a higher velocity, would catch up with the synchronous particle. Now that we understand dispersion, we begin to realize that the situation may be more complicated. Giving the errant particle more energy will speed it up but may also send it on an orbit of larger radius.

The path length that the errant particle must travel around the machine, or more correctly, the change in path length with momentum, must depend upon the dispersion function. The closed orbit will have a mean radius

$$R = R_0 + \bar{D}\frac{\Delta p}{p}.$$

It can be argued that, close to the velocity of light where acceleration can increase momentum but not velocity, the longer path length will more than cancel the small effect of velocity and the particle, instead of catching up with its synchronous partner, will arrive even later than it did on the previous turn. This seems to defeat the whole idea of phase stability.

We shall see that, depending on how the synchrotron is designed and which particles it accelerates, there can be a certain energy where our initial ideas of phase stability break down. This is called the transition energy. There is also a way of ensuring stability above transition.

5.3 Transition energy

The crucial factor in resolving the question of velocity versus path length is the way in which the revolution time (or its reciprocal, the revolution frequency) varies as the particle is given extra acceleration. The revolution frequency is

$$f = \frac{\beta c}{2\pi R} \quad \left(\beta = \frac{v}{c}\right).$$

We see that f depends on two momentum-dependent variables, Einstein's β and R, the mean radius. The penultimate equation gives the change in the radius. (Incidentally, both of the above equations are only strictly valid if the average guide field in the synchrotron is constant over many turns but we allow ourselves this assumption for the moment.) The momentum dependence of β is determined by

$$pc = \frac{E_0\beta}{\sqrt{1-\beta^2}}.$$

We can best express the rate of 'catching up' upon which phase stability depends by a 'slip factor' η, which is defined as logarithmic differential of frequency as a function of momentum. The procedure of partial derivatives implies that there must be two terms corresponding to the two variables which are momentum dependent. Hence,

$$\eta_{\text{rf}} = \frac{\Delta f/f}{\Delta p/p} = \frac{p}{\beta}\frac{d\beta}{dp} - \frac{p}{R}\frac{dR}{dp} = \frac{1}{\gamma^2} - \frac{\bar{D}}{R_0}.$$

The first term on the right-hand side describes the increase in speed with p and the other, which is negative, how the path to be travelled increases with p.

Only the first of the two terms on the right-hand side varies as particles are accelerated. At low energy, when $\gamma = 1$, this is largest and η is usually positive. But, since $\gamma = E/E_0$, the first term shrinks with E and becomes smaller than the second at high energy so that η changes sign from positive to negative. There is a certain energy, the transition energy, at which η is momentarily zero. At this transition energy the value of γ satisfies

$$\frac{1}{\gamma_{\text{tr}}^2} = \frac{\bar{D}}{R}.$$

In a conventional proton synchrotron, transition is usually encountered midway through the acceleration cycle and can only be avoided with some ingenuity in the design of the lattice. This was a considerable worry to the designers of the CERN proton synchrotron (CPS) and AGS. These were the first proton synchrotrons of high enough energy to encounter this problem during acceleration and, using the new alternating-gradient focusing, had small values of D. Fortunately, it was realized that one could almost instantaneously change the phase of the voltage wave in the r.f. cavities to be falling rather than rising at the moment of the synchronous particle's arrival (Fig. 5.9). Above transition and with such a reversed slope, particles arriving late are given less than their ration of energy and take an inner circular path—a short cut—to arrive earlier next time.

There was still a worry about changing the phase rapidly enough and theorists speculated about the regions on either side of this change from one phase to another where they feared there would be insufficient phase stability of either kind. It was therefore a great relief to the CPS builders to find that when acceleration was first tried a simple low-level r.f. circuit, built by W. Schnell inside a

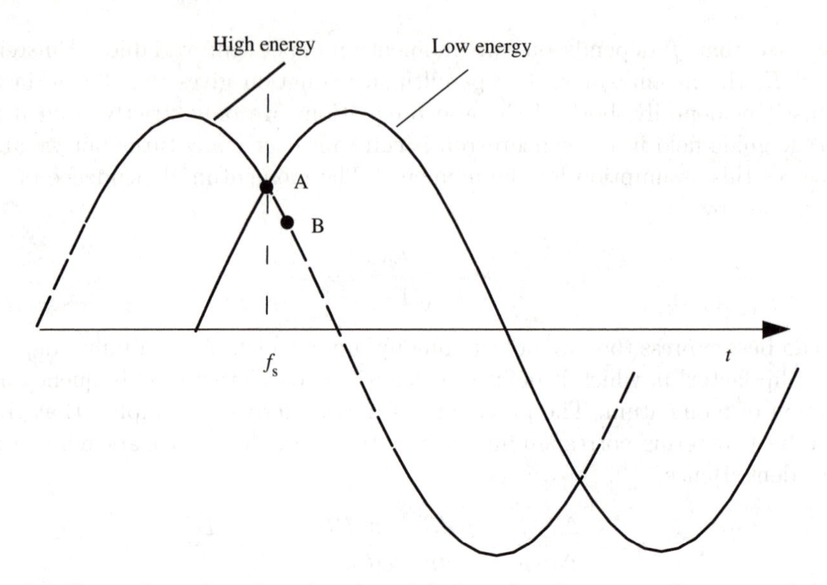

Fig. 5.9 Shows how changing the phase of the r.f. voltage waveform can give the lagging particle, B, less energy rather than more and can lead to stability above transition.

treacle tin to switch the phase, took the machine smoothly through this transition with hardly a proton lost.

Electron machines are fortunate in that γ, being 2000 times higher, ensures that the first term may be neglected so that they always operate above transition.

5.4 Synchrotron motion

We now come to the core of our discussion of longitudinal motion. If we consider the motion of a particle on the linear part of the voltage wave of an r.f. cavity, it is not difficult to imagine that it approximates rather closely to a harmonic oscillator. The motion becomes more complicated when the particle's amplitude is larger and for part of its motion it finds itself over the crest of the wave, but first let us focus on a small-amplitude solution.

It is not hard to deduce from special relativity that the momentum may be written as

$$p = m_0 c(\beta\gamma).$$

We shall therefore use $\Delta(\beta\gamma)$ as the momentum coordinate in longitudinal phase space. The other coordinate is the particle's arrival phase, ϕ, with respect to the zero-crossing of the r.f. voltage at the cavity. Let us consider the simplest case of a small oscillation in a stationary bucket, $\phi_S = 0$, when the particle is not being accelerated.

A particle with a small phase error will describe an ellipse in phase space which one may write parametrically as

$$\Delta(\beta\gamma) = \widehat{\Delta(\beta\gamma)} \sin 2\pi f_s t,$$
$$\phi = \hat{\phi} \cos 2\pi f_s t,$$

where f_s is the frequency of execution of these oscillations in phase, which we call the synchrotron frequency.

In order to find the differential equation corresponding to this motion, we must first remember that the angular frequency $2\pi f$ of an oscillator is nothing other than $\dot{\phi}$, the rate of change of phase or, to be exact, $-\dot{\phi}$. (The negative sign stems from the fact that ϕ is a phase lag.) For example, a 5 Hz frequency changes phase at a rate of $10\pi \,\text{rad s}^{-1}$. We may therefore relate the rate of change in arrival phase to the difference in revolution frequency of the particle, compared to that of the synchronous particle:

$$\dot{\phi} = -2\pi h[f(\Delta\beta\gamma) - f(0)] = -2\pi h \Delta f.$$

Note that we have multiplied by h, the harmonic number of the r.f., since ϕ is the phase angle of the r.f. swing while $f(\Delta\beta\gamma)$ refers to the revolution frequency. Here we can use the definition of the slip factor η and then simply use some standard relativistic relations to end up with Δf as a function of ΔE, the energy defect with respect to the synchronous particle:

$$\Delta f = \eta f \frac{\Delta p}{p} = \eta f \frac{\Delta(\beta\gamma)}{(\beta\gamma)} = \frac{\eta f}{\beta^2} \frac{\Delta\gamma}{\gamma} = \frac{\eta f}{E_0 \beta^2 \gamma} \Delta E,$$

where E_0 is the particle's rest energy.

Having made the substitution, we differentiate once more to obtain a second-order differential equation which, we hope, resembles a simple oscillator:

$$\ddot{\phi} = -\frac{2\pi h \eta f}{E_0 \beta^2 \gamma} (\Delta \dot{E}).$$

We recall from the arguments of phase stability that per turn the extra energy given to a particle whose arrival phase is ϕ will be

$$\Delta E = eV_0(\sin\phi - \sin\phi_s)$$

and that the rate of change of energy will therefore be ΔE times f, the revolution frequency. We can write

$$\ddot{\phi} = -\frac{2\pi e V_0 \eta h f^2}{E_0 \beta^2 \gamma} (\sin\phi - \sin\phi_s).$$

This is actually a fundamental and exact description of the motion provided the parameters change slowly (the adiabatic assumption). The practice now is to simply integrate this to find its solution numerically. But to see an analytic solution for small amplitudes we set $\phi_s = 0$ and $\phi = \sin\phi$:

$$\ddot{\phi} + \frac{2\pi e V_0 \eta h f^2}{E_0 \beta^2 \gamma}\phi = 0.$$

Clearly, the frequency of these synchrotron oscillations in longitudinal phase space is

$$f_s = \sqrt{\frac{|\eta| h e V_0}{2\pi E_0 \beta^2 \gamma}}\, f,$$

or, writing $f_{rf} = hf$, we could also express it as

$$f_s = \sqrt{\frac{|\eta| e V_0}{2\pi E_0 \beta^2 \gamma h}}\, f_{rf}.$$

We can also define a synchrotron tune, Q_S, as the number of such oscillations per revolution of the machine. This is analogous to Q in transverse phase space:

$$Q_s = \frac{f_s}{f} = \sqrt{\frac{|\eta| e h V_0}{2\pi E_0 \beta^2 \gamma}}.$$

In most machines Q_s is of the order of 10% of the revolution frequency or less. It sweeps down to zero at γ transition where η is zero and then rises again. In large proton machines it can be in the 0–100 Hz region and, but for the vacuum, one might hear it! It can cause trouble when it crosses the harmonics of 50 Hz, which occur in the power supply ripple, and the radial servo loop for r.f. cavity voltage and frequency can cause a resonance.

Close to γ_{tr} we cannot strictly assume that β, γ, η, and f vary slowly in comparison with the synchrotron oscillation which this equation describes. Hence, we should have been more careful about obtaining the second derivative above and the reader is advised to use a more exact form of the equation of motion and approximate only when it seems that this is justified. (The texts which discuss adiabaticity will prove comforting if rather tedious reading.) This more exact form of the equation of motion is

$$\frac{d}{dt}\left[\frac{E_0 \beta^2 \gamma \dot{\phi}}{2\pi \eta h f^2}\right] + eV_0(\sin\phi - \sin\phi_s) = 0.$$

In a stationary bucket, when $\phi_s = 0$, this exact differential equation for large amplitude motion is identical to that for a rigid pendulum. Such a pendulum—a mass suspended on a weightless rigid rod—displaced by angle θ will execute oscillations according to

$$l\frac{d^2\theta}{dt^2} + g\sin\theta = 0.$$

There is an extra term, $\sin\phi_S$, on the right-hand side of the synchrotron equation which is not there in the pendulum case but it could be introduced for the pendulum too by using a magnetic 'bob' and biasing its equilibrium position to one side by placing a magnetic field in an asymmetric position.

One of the stumbling blocks to the beginner is the complicated shape of the bucket (Fig. 5.2), which leads to a trivial but very lengthy calculation of the area of the bucket as a function of ϕ_S, V, and η. Brück (1966) gives a good explanation of how one does this analytically. Nowadays we write numerical integration routines to do this or track particles with the simple algorithm $\Delta E = V_0(\sin\phi - \sin\phi_S)$ for each turn.

Exercises

5.1 Write down the expression for total energy in terms of rest energy and momentum (pc) and thence derive an expression for γ as a function only of β.

5.2 Solve to give β as a function of γ and then as a function of E and E_0.

5.3 Show that $pc = E_0(\beta\gamma)$.

5.4 Establish the following relationships:

$$dE = \beta c\, dp, \qquad \frac{dE}{E} = \beta^2 \frac{dp}{p},$$

$$\frac{df}{f} = \left(\frac{1}{\gamma^2} - \alpha\right)\frac{dp}{p}, \quad \text{where } \alpha = \frac{dR/R}{dp/p},$$

f being the revolution frequency in a synchrotron in which the momentum compaction is α. R is the physical radius (circumference $= 2\pi R$).

5.5 A 10 GeV (kinetic energy) synchrotron, has a magnetic field which rises to 1.5 T in 1 s. Given that the mass of the proton is 0.9383 GeV: What is the momentum at 1.5 T? What is $B\rho$? If 2/3 of the circumference is bending magnets, what are ρ and R, the mean radius? What is the revolution frequency at 10 GeV? What is the revolution frequency at 1 GeV? Assuming the revolution frequency at 1 GeV, calculate the voltage per turn necessary to provide a linear rate of rise of 9 GeV/s. If $\sin\phi_S = \sin 45°$, what is the peak voltage necessary in the cavity?

5.6 If the mean dispersion around the ring is 9 m, what is (a) γ_{tr}? (b) momentum at transition? (c) η at 1 GeV and at 10 GeV?

5.7 If the harmonic number is 10, what is the synchrotron frequency at 1 GeV?

5.8 Write a small computer program to plot f_s in steps of 0.05 GeV from 1 to 2 GeV.

6 Imperfections and multipoles

6.1 A simplified treatment of betatron motion

The earlier chapters on focusing treat betatron motion in a rather rigorous way. However, some readers who are new to the field might find the following sections too confusing if we carry through all the terms from the rigorous theory into a study of imperfections, whose effect we can anyway only estimate. We shall therefore introduce two models of the motion which are approximate but graphic.

6.1.1 The circle diagram

In Fig. 6.1(b) we see the phase-space ellipse of a beam, plotted at a point near an F quadrupole where the amplitude function β is large. The ellipse will be very wide and not very high in divergence angle and it is in such positions that a small angular kick has the greatest effect on the beam. Imagine, for example, how little angular displacement is needed to move the ellipse by its own height and increase the emittance by a factor of 2. Thus in a machine with a FODO

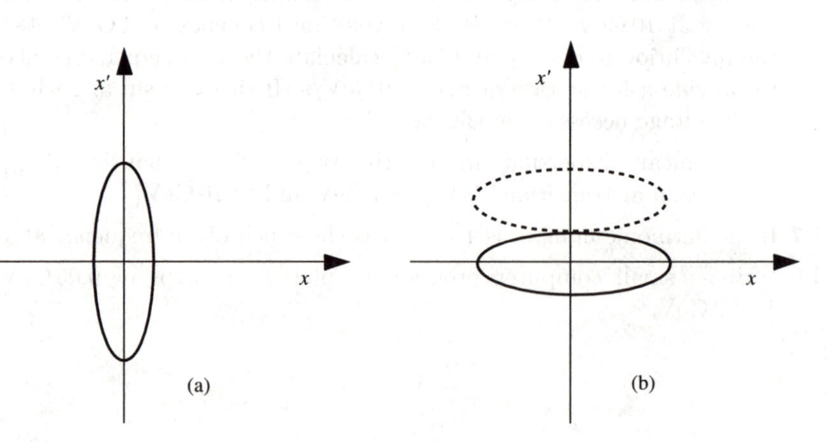

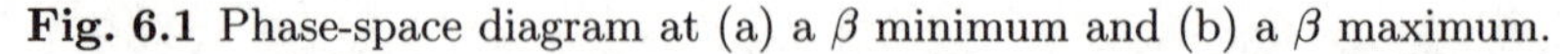

Fig. 6.1 Phase-space diagram at (a) a β minimum and (b) a β maximum.

lattice, most of the damage is done to the beam by bad fields which cause angular kicks near the F quadrupoles.

So predominant are the effects of perturbations near $\hat{\beta}$ positions that one can often do quite good 'back of the envelope' calculations by closing one's eyes to what happens to the protons in between the F quadrupoles. At F quadrupoles the ellipse always looks the same, that is, upright, with displacement and divergence semi-axes

$$\sqrt{\beta\varepsilon} \quad \text{and} \quad \sqrt{\frac{\varepsilon}{\beta}},$$

respectively. This can be reduced to a circle of radius $\sqrt{\beta\varepsilon}$, by using the new coordinates

$$y = y, \qquad p = \beta y'.$$

Note that p is not related to our earlier use of the symbol while β is a Twiss parameter again.

If the machine has, for example, 108 periods and a Q of 27.6, a proton will advance in phase by $2\pi Q/108$ from one period to the next; this is just the angle subtended at the centre of the circle multiplied by Q. After one turn of the machine, it has made 27 revolutions of the circle plus an angle of 2π multiplied by the fractional part of Q, see Fig. 6.2.

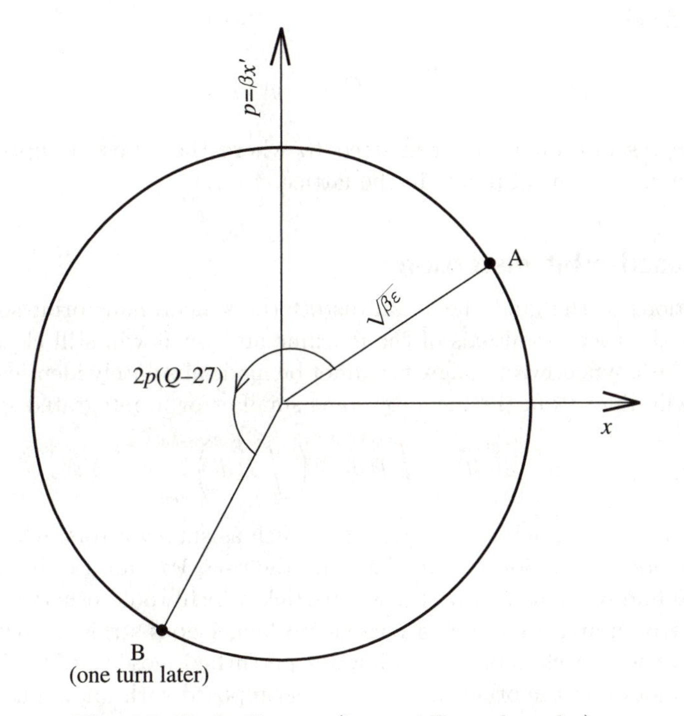

Fig. 6.2 Circle diagram (locus at F quadrupoles).

6.1.2 The (η, ψ) description of alternating-gradient focusing

This normalization of the phase space can be done in a more rigorous way by choosing new variables (η, ψ) which transform the distortion of the phase and amplitude so that the motion becomes that of a harmonic oscillator. We must, of course, transform back again to see physical displacements, but the mathematics in the new coordinates becomes more transparent. We will not dwell on how this transformation—sometimes called Floquet's transformation—was found, but just state it. The transformed coordinates are related to the old by

$$\eta = \beta^{-1/2} y,$$

$$\psi = \int \frac{ds}{Q\beta}, \qquad g(\psi) = Q^2 \beta^{3/2} F(s),$$

where ψ advances by 2π every revolution. It coincides with θ at each location where β is maximum or minimum and does not depart very much from θ in between. The function $g(\psi)$ is the transformation of an azimuthal perturbation of the guide field,

$$F(s) = \frac{\Delta B(s)}{B\rho},$$

which, in the ideal case, is everywhere zero. With this transformation, Hill's equation becomes the harmonic oscillator equation with a driving term on the right-hand side:

$$\frac{d^2\eta}{d\psi^2} + Q^2\eta = g(\psi).$$

This representation is not restricted to where the ellipse is upright but is rigorous and valid at all points in the lattice.

6.2 Closed-orbit distortion

Imperfections in the guide field can distort the synchronous orbit so that it is no longer the theoretical axis of the machine and yet it will still close on itself. Even the best synchrotron magnets cannot be made absolutely identical and each magnet will differ from the mean by some small error in integrated strength:

$$\delta(Bl) = \int B\,dl - \left(\int B\,dl\right)_{\text{ideal}}$$

These and other machine imperfections, such as survey errors, which give rise to field errors, are randomly spread around the ring. We can use the (η, ψ) coordinates to find out how this perturbs a particle which would otherwise have had zero betatron amplitude. Such a particle no longer goes straight down the centre of the vacuum chamber but follows a perturbed closed orbit. The normal betatron motion of the other protons is superimposed with undiminished amplitude about this distorted closed orbit. One of the most important considerations in designing and later in setting up a machine is to keep the distortion of this closed orbit to a minimum because it eats up available machine aperture. It used to be conventional wisdom in designing a machine to make sure that the vacuum chamber would accommodate twice the expectation value of distortion. The probability of no particles making the first turn was thus reduced to a mere 2%. Designers of modern large machines cannot afford this luxury and must rely on closed-orbit steering with correcting dipole magnets to thread the first turn. Once a few turns are circulating round the machine these correctors can completely compensate the distortion. However, in order to design these corrector magnets we must estimate the magnitude of the distortion to be corrected. As a first step, let us consider the effect on the orbit of a small additional dipole located at a position where $\beta = \beta_K$ and observed at another position.

6.2.1 Distortion due to a single dipole

Suppose we gradually turn on a short dipole (we shall assume it is a delta function in s) which makes a growing angular 'kink' in divergence in the trajectory of every turn (Fig. 6.3):

$$\delta y' = \frac{\delta(Bl)}{B\rho}.$$

The closed-orbit trajectory is perturbed by a cusp while, elsewhere, the trajectory must obey the laws of betatron motion which, expressed in (η, ψ) coordinates, are

$$\frac{d^2\eta}{d\psi^2} + Q^2\eta = 0,$$

$$\eta = \eta_0 \cos(Q\psi + \lambda).$$

In Fig. 6.4 we have chosen the $\psi = 0$ origin to be diametrically opposite to the kick and, by symmetry, $\lambda = 0$ and the 'orbit' starts and ends as a cosine function.

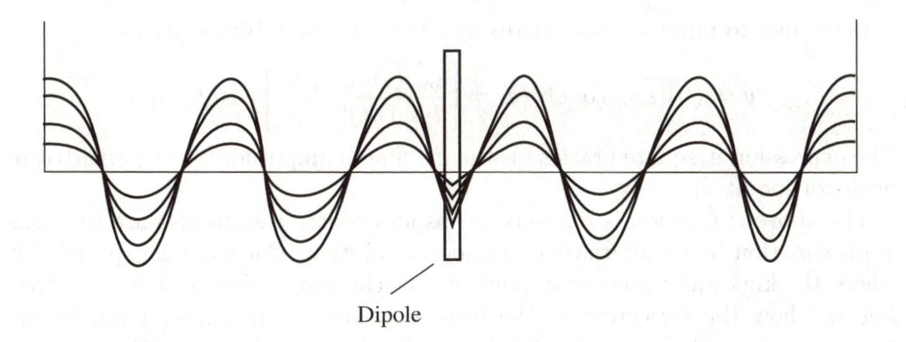

Fig. 6.3 Closed-orbit distortion as a dipole is slowly switched on.

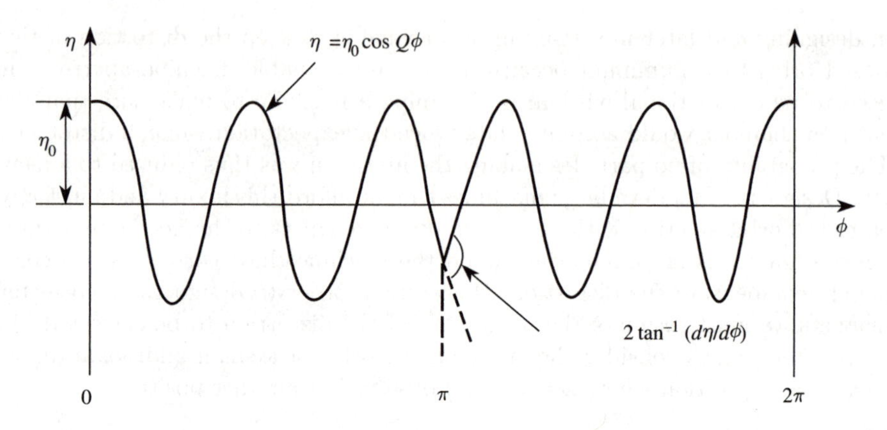

Fig. 6.4 Tracing the closed orbit for one turn in (η, ψ) space with a single kick at $\psi = \pi$. (The Q value is about 4.6.)

We only consider a trajectory which is closed, and continuity therefore demands that the kick $\delta y'$ matches the change in slope at $\psi = \pi$, the location of the dipole.

Differentiating the orbit equation, we have

$$\frac{d\eta}{d\psi} = -\eta_0 Q \sin Q\psi = -\eta_0 Q \sin Q\pi.$$

To relate this to the real kick we use

$$\frac{d\psi}{ds} = \frac{1}{Q\beta_{\mathrm{K}}}, \qquad \frac{dy}{ds} = \sqrt{\beta_{\mathrm{K}}}\frac{d\eta}{ds};$$

therefore,

$$\frac{\delta y'}{2} = \frac{\delta(Bl)}{2B\rho} = \frac{dy}{ds} = \sqrt{\beta_{\mathrm{K}}}\frac{d\eta}{d\psi}\frac{d\psi}{ds} = \frac{\eta_0}{\sqrt{\beta_{\mathrm{K}}}}\sin \pi Q,$$

$$\eta_0 = \frac{\sqrt{\beta_{\mathrm{K}}}}{2|\sin \pi Q|}\delta y'.$$

Returning to physical coordinates we can write the orbit's equation

$$y = \sqrt{\beta(s)}\eta_0 \cos Q\psi(s) = \left[\frac{\sqrt{\beta(s)\beta_{\mathrm{K}}}}{2|\sin \pi Q|}\frac{\delta(\beta l)}{B\rho}\right]\cos Q\psi(s).$$

The expression in square brackets is the maximum amplitude of the perturbation or distortion at $\beta(s)$.

The above expression is rigorous but as an exercise we can also use the circle approximation as an alternative method of solution. Consider the special case where the kink and observation point are at the same value of β. We see from Fig. 6.5 how the trajectory of the distortion appears in phase space. Simple trigonometric analysis reveals that the amplitude a is just the quantity in square brackets.

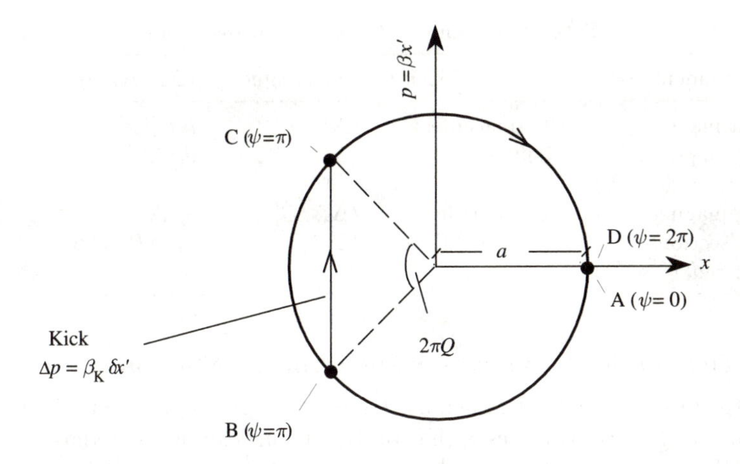

Fig. 6.5 Tracing a closed orbit for one turn in the circle diagram with a single kick. The path is ABCD.

6.2.2 Effect of many errors

In estimating the effect of a random distribution of dipole errors, we must take the r.m.s. average, weighted according to the β_i values, over all of the kicks $\delta y_i'$ from the N magnets in the ring. The expectation value of the amplitude is

$$\langle y(s) \rangle = \frac{\sqrt{\beta(s)}}{2\sqrt{2}\sin \pi Q}\sqrt{\sum_i \beta_i \delta y_i'^2}$$

$$= \frac{\sqrt{\beta(s)\bar{\beta}}}{2\sqrt{2}\ \sin \pi Q}\sqrt{N}\frac{(\Delta Bl)_{\text{rms}}}{B\rho}.$$

The factor $\sqrt{2}$ comes from averaging over all the phases of distortion produced.

6.2.3 Sources of distortion

The principal imperfections in a synchrotron causing orbit distortion are shown in Table 6.1. The first line in the table lists the random variations in the position of quadrupole magnets with respect to their ideal location. A small transverse displacement of a quadrupole gives an effective dipole perturbation, $kl\Delta y$. The tilt of bending magnets in the second line causes a small resultant dipole in the horizontal direction, which deflects vertically. Random errors in magnet gap, length or in the coercivity of the steel yoke which determines remanent field, contribute to the third line. Both remanent and stray fields in straight sections tend to be constant and their effect scales as $1/B$ as the machine pulses. Their effect should therefore be evaluated at injection where it is worst. In a modern superconducting machine the persistent current fields play the same role as remanent effects.

Table 6.1 Sources of closed-orbit distortion

Type of element	Source of kick	r.m.s. value	$\langle \Delta Bl/(B\rho)\rangle_{\mathrm{rms}}$	Plane
Gradient magnet	Displacement	$\langle \Delta y\rangle$	$k_i l_i \langle \Delta y\rangle$	x, z
Bending magnet (bending angle $= \theta_i$)	Tilt	$\langle \Delta\rangle$	$\theta_i \langle \Delta\rangle$	z
Bending magnet	Field error	$\langle \Delta B/B\rangle$	$\theta_i \langle \Delta B/B\rangle$	x
Straight sections (length $= d_i$)	Stray field	$\langle \Delta B_{\mathrm{s}}\rangle$	$d_i \langle \Delta B_{\mathrm{s}}\rangle/(B\rho)_{\mathrm{inj}}$	x, z

6.2.4 The Fourier harmonics of the error distribution

One of the physical insights gained by reducing the problem to that of a harmonic oscillator in (η, ψ) coordinates is that perturbations can be treated as the driving term of the oscillator. They may be broken down into their Fourier components and the whole problem seen as the forced oscillations of a pendulum. The driving term is put on the right-hand side of Hill's equation:

$$\frac{d^2\eta}{d\psi^2} + Q^2\eta = Q^2 \sum_{k=1}^{\infty} f_k e^{ik\psi} = Q^2 \beta^{3/2} F(s),$$

where $F(s)$ is the azimuthal pattern of the perturbation $\Delta B/(B\rho)$; and $Q^2\beta^{3/2}$ comes from the transformation from physical coordinates to (η, ψ) coordinates.

The Fourier amplitudes are defined by

$$f(\psi) = \beta^{3/2} F(s) = \sum_k f_k e^{ik\psi},$$

where

$$f_k = \frac{1}{2\pi} \int_0^{2\pi} f(\psi) e^{-ik\psi}\, d\psi = \frac{1}{2\pi Q} \oint \beta^{1/2} F(s) e^{-ik\psi}\, ds.$$

We can then solve Hill's equation as

$$\eta = \sum_{k=1}^{\infty} \frac{Q^2 f_k}{Q^2 - k^2} e^{ik\psi} \quad \text{(or its real part)}.$$

But be careful! Before doing the Fourier analysis, ΔB must be multiplied by $\beta^{1/2}$ if the physical variable s is chosen as an independent variable, or $\beta^{3/2}$ if ψ, the transformed phase, is used and these are themselves functions of s.

Looking carefully at the above expression, we see that this differs from the general solutions

$$\eta = \eta_0 e^{\pm iQ\psi},$$

which describe the betatron motion about the equilibrium orbit, because the wavenumber is an integer k. In fact this *closed* orbit is a particular solution of

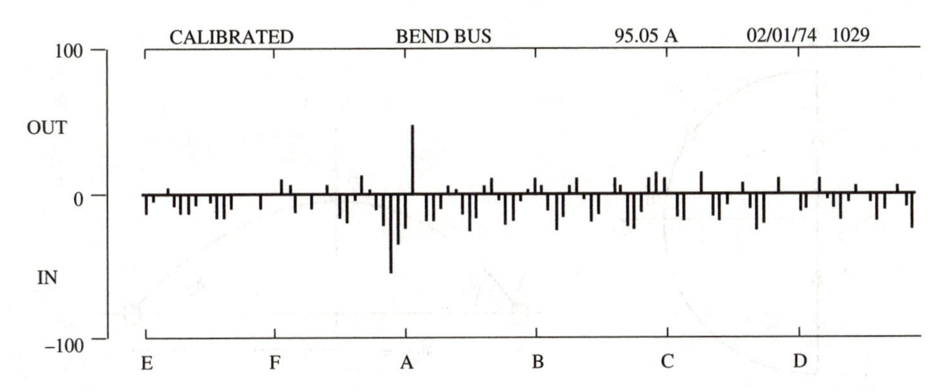

Fig. 6.6 FNAL main ring electrostatic pickups show closed orbit around the ring ($Q \equiv 19.2$).

Hill's differential equation, to which we must add the general solutions which describe the betatron oscillations about it.

The function $Q^2/(Q^2 - k^2)$ is called *the magnification factor* for a particular Fourier component of ΔB. It rises steeply when the wavenumber k is close to Q, and the effect of the two Fourier components in the random error pattern which have k values adjacent to Q accounts for about 60% of the total distortion due to all random errors. Figure 6.6 shows an uncorrected closed-orbit pattern from electrostatic pickups in the old Fermilab Main Ring, whose Q is between 19 and 20. The pattern shows strong components with these wavenumbers. If Q is deliberately tuned to an integer k, the magnification factor is infinite and errors of that frequency make the proton walk out of the machine. This is in fact an integer resonance driven by dipole errors.

6.2.5 Closed-orbit bumps

It is often important to make a closed-orbit bump deliberately at one part of the circumference without affecting the central orbit elsewhere. A typical example of this is to make the beam ride close to an extraction septum or within the narrow jaws of an extraction kicker magnet placed just outside the normal acceptance of the ring.

If one is lucky enough to find two positions for small dipole deflecting magnets spaced by π in betatron phase and centred about the place where the bump is required, the solution is very simple (Fig. 6.7). The distortion produced is

$$y(s) = \delta\sqrt{\beta(s)\beta_{\mathrm{k}}}\sin(\phi - \phi_0),$$

where $\beta(s)$ is the beta function and β_{k} at the deflector and

$$\delta = \frac{\Delta(Bl)}{B\rho}.$$

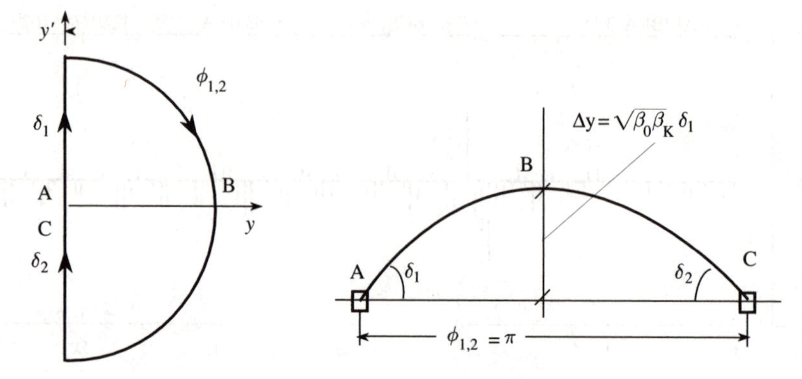

Fig. 6.7 An exact half-wave bump using two dipoles.

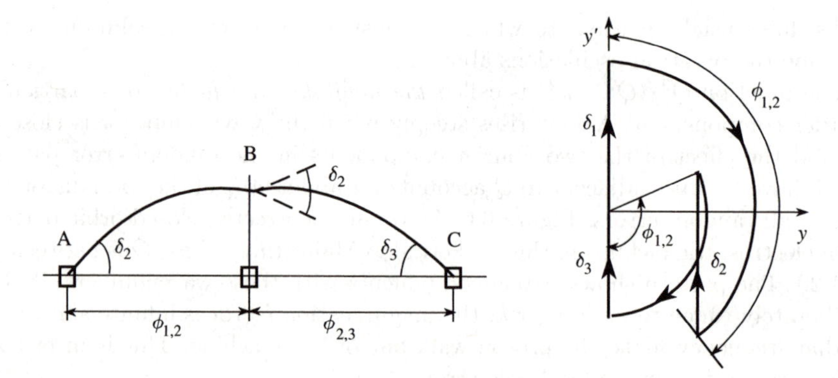

Fig. 6.8 A general bump using three dipoles.

Here ϕ is the betatron phase and ϕ_0 at the dipole. This half-wave bump also has a very simple configuration in normalized phase space. We see that the central orbit (at the origin) is not disturbed elsewhere. Note that the magnitude of the bump is not only proportional to the root of the local $\beta(s)$ but is proportional to β_{k}, the value of β where the kick due to the dipole is applied. Since β is largest at F quadrupoles, this is clearly where one should locate dipole bumpers.

Very often, F quadrupoles are not π apart in phase but separated by some other phase interval, say $2\pi/3$. This means the possible locations for dipoles are slightly less effective ($\sin\pi/3 = \sqrt{3}/2$ in our example) and we must introduce a third dipole to form a triad if the orbit perturbation is to be confined to within the bump. The third dipole is best located near the peak of the bump.

Figure 6.8 shows how the three bumps add up in normalized phase space. The case illustrated is the general one with dipoles of different strengths and spaced differently in phase. In order to find an exact solution to the problem of a triad bump we use the matrix which transforms a point in phase space from one location to another:

$$\begin{pmatrix} y \\ y' \end{pmatrix} =$$

$$\begin{pmatrix} (\sqrt{\beta}/\sqrt{\beta_0})(\cos\Delta\phi + \alpha_0 \sin\Delta\phi) & \sqrt{\beta_0\beta}\sin\Delta\phi \\ \frac{-1}{\sqrt{\beta_0\beta}}\{(\alpha-\alpha_0)\cos\Delta\phi + (1+\alpha\alpha_0)\sin\Delta\phi\} & \frac{\sqrt{\beta_0}}{\sqrt{\beta}}(\cos\Delta\phi - \alpha\sin\Delta\phi) \end{pmatrix} \cdot \begin{pmatrix} y_0 \\ y' \end{pmatrix}$$

It is the element which links y to y_0' which describes the trajectory. Following the kick δ_1,

$$y_2 = \delta_1 \sqrt{\beta_2\beta_1}\sin(\phi_2 - \phi_1).$$

The same argument can be used to describe the trajectory working back from δ_3,

$$y_2 = -\delta_3 \sqrt{\beta_2\beta_3}\sin(\phi_2 - \phi_3).$$

The kick δ_2 must be the change in the derivative:

$$\delta_2 = \delta_1 \frac{\beta_1}{\beta_2}\cos(\phi_2 - \phi_1) + \delta_3 \frac{\beta_3}{\beta_2}\cos(\phi_3 - \phi_2).$$

We can rewrite these relations as

$$\sqrt{\beta_1}\delta_1 \sin\phi_{12} = \sqrt{\beta_3}\delta_3 \sin\phi_{23},$$

$$\sqrt{\beta_2}\delta_2 = \sqrt{\beta_1}\delta_1 \cos\phi_{12} + \sqrt{\beta_3}\delta_3 \cos\phi_{23},$$

where

$$\phi_{ij} = \phi_i - \phi_j.$$

These relations are those that apply to a triangle of sides $\delta\sqrt{\beta}$, and angles ϕ, which can be solved by the well-known symmetric relation

$$\frac{\delta_1\sqrt{\beta_1}}{\sin\phi_{23}} = \frac{\delta_2\sqrt{\beta_2}}{\sin\phi_{31}} = \frac{\delta_3\sqrt{\beta_3}}{\sin\phi_{12}}.$$

6.2.6 The measurement and correction of closed orbits

Electrostatic plates with diagonal slots (Fig. 6.9) are commonly used to measure the transverse position of a bunched beam. We have seen that the predominant harmonic in the uncorrected orbit is close to Q and, to establish its amplitude and phase, one really needs four pickups per wavelength. Given the present fashion for FODO lattices with about 90° per cell and the need to measure in both planes, the final solution is usually to place one pickup at each quadrupole. The ones at F quadrupoles, where β is large horizontally, are the most accurate for the horizontal plane, while others at D quadrupoles are best for the vertical correction.

Similar arguments lead us to have one horizontally defecting dipole at each F quadrupole, where β_x is large, and one vertically deflecting dipole at each D quadrupole, where β_z is large.

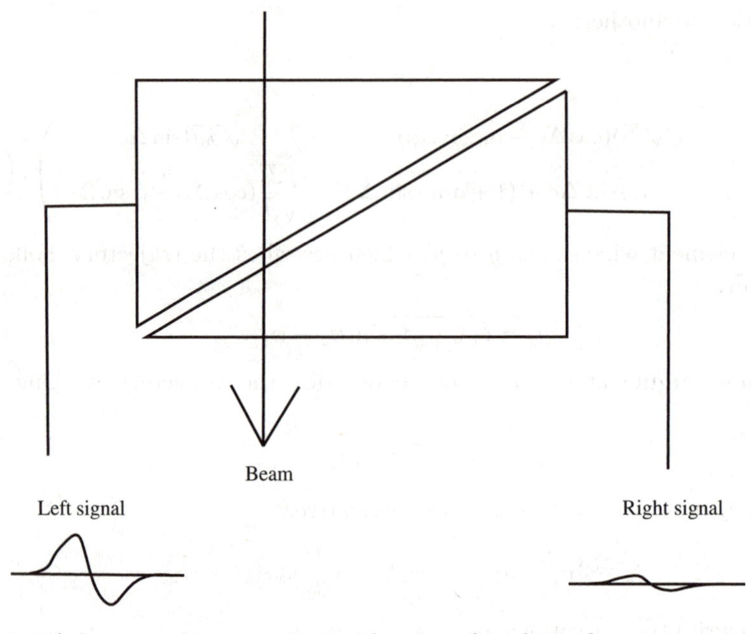

Fig. 6.9 A beam position monitor with triangular electrodes gives an asymmetric signal.

Clearly, many of these correcting dipoles would be unnecessary if only the two principal harmonics are to be corrected and one method of correcting orbit distortion is to apply a pattern of dipole correctors which excite an equal and opposite Fourier component of error at an integer close to Q. For example, one can excite an odd harmonic such as Fermilab's 19th with a single dipole, though a pair in opposition are necessary if their average bending effect, which influences the nominal momentum, is to cancel. Also, a further pair are needed in quadrature to adjust the phase, making four per harmonic and per plane. Another set will be needed for even harmonics. This is shown in Fig. 6.10. However, instructive though this may be, it leaves about 40% of the distortion uncorrected. It is sometimes used as a last resort to find out what the integer part of Q really is. We mention this because such harmonic techniques are used for the correction of non-linear resonances, where one need only correct the principal harmonics.

The more generally used method of correcting orbit distortion consists in applying a set of superposed beam bumps, each formed along the lines calculated above. Triads of correctors are selected from the regular pattern around the lattice and each calculated to compensate the measured orbit position at its centre. Given the power of modern computers, this kind of correction can be calculated and applied all round the ring instantly.

Some machines do not have dipole correctors which are sufficiently strong to correct an orbit at their top energy and quadrupole magnets must then be

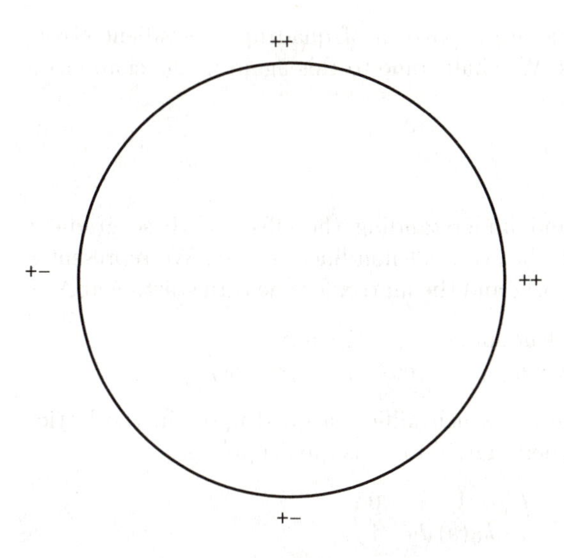

Fig. 6.10 Diagram showing the sign of correction dipoles necessary to excite or compensate even or odd Fourier components of distortion around the ring of a synchrotron.

displaced upwards or sideways a distance Δy to apply an effective dipole

$$\frac{\Delta(Bl)}{B\rho} = k\Delta y.$$

Such displacements are tedious to apply to all quadrupoles in large machines and indeed the accumulated effect of errors in moving so many quadrupoles by a few tenths of a millimetre might even make the orbit worse. An alternative to moving all quadrupoles (or powering all dipoles) is to select those which are most effective in correcting the orbit. Stored in the computer is a large matrix **G** with as many rows as pickups and as many columns as correctors. Each term describes the effect y_i of a corrector Δ_j at the ith pickup—the result of a beam optics calculation or even a measurement (Autin and Bryant 1971):

$$\begin{pmatrix} y_1 \\ y_2 \\ y_3 \\ \vdots \\ y_n \end{pmatrix} = \begin{pmatrix} & & & \\ & G_{pq} & & \\ & & & \end{pmatrix} \begin{pmatrix} \Delta_1 \\ \Delta_2 \\ \Delta_3 \\ \vdots \\ \Delta_n \end{pmatrix}.$$

The matrix may be inverted to calculate the corrections for any set of measurements **Y**. But since we first want to select the most effective correctors, we include only those correctors which correspond to large matrix elements. The method also has the advantage that it can be used if one or more of the pickups are out of action.

One last comment is that while orbit correction was originally invented to economize on magnet aperture, precise correction is now considered essential to reduce the effects of non-linearities in the dynamics of synchrotrons. For example, uncorrected orbit distortion in the sextupoles, which most machines have for correcting chromaticity, will generate a pattern of quadrupole gradient errors and drive half-integer stopbands. We shall come to this again in the chapter on non-linearities.

6.3 Gradient errors

Quadrupoles also have errors and understanding the effect of these gradient errors is a useful preparation for the study of non-linear errors. We represent a ring of magnets as a circle in Fig. 6.11 and the matrix for one turn starting at A as

$$M_0(s) = \begin{pmatrix} \cos\varphi_0 + a_0 \sin\varphi_0 & \beta_0 \sin\varphi_0 \\ -\gamma_0 \sin\varphi_0 & \cos\varphi_0 - a_0 \sin\varphi_0 \end{pmatrix}.$$

Now consider a small gradient error which afflicts a quadrupole in the lattice between B and A. The unperturbed matrix for this quadrupole is

$$m_0 = \begin{pmatrix} 1 & 0 \\ -k_0(s)\,ds & 1 \end{pmatrix}$$

and, when perturbed, the quadrupole matrix

$$m = \begin{pmatrix} 1 & 0 \\ -[k_0(s) + \delta k(s)]\,ds & 1 \end{pmatrix}.$$

The unperturbed transfer matrix for the whole machine includes m_0. To find the perturbed transfer matrix we make a turn, backtrack through the

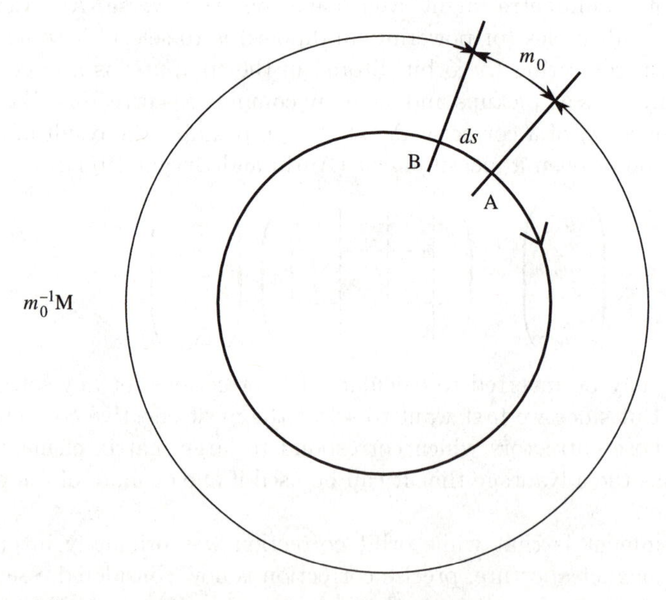

Fig. 6.11 Matrix representation of a small quadrupole, m_0, subject to an error which is a component of the matrix for the whole ring, M.

small unperturbed quadrupole (m_0^{-1}), and then proceed through the perturbed quadrupole (m). Translating this into matrix algebra, we have

$$M(s) = m m_0^{-1} M_0.$$

Now

$$m m_0^{-1} = \begin{pmatrix} 1 & 0 \\ -\delta k(s)\, ds & 1 \end{pmatrix}.$$

So,

$$M = \begin{pmatrix} \cos\phi_0 + \alpha_0 \sin\phi_0 & \beta_0 \sin\phi_0 \\ -\delta k(s)\, ds(\cos\phi_0 + \alpha_0 \sin\phi_0) - \gamma \sin\phi_0 & -\delta k(s)\, ds \beta_0 \sin\phi_0 + \cos\phi_0 - \alpha_0 \sin\phi_0 \end{pmatrix}.$$

Now $\mathrm{Tr}\, M = 2\cos\phi$. So the change in $\cos\phi$ is

$$\Delta(\cos\phi) = -\Delta\phi \sin\phi_0 = \frac{\sin\phi_0}{2} \beta_0(s)\, \delta k(s)\, ds,$$

$$2\pi\Delta Q = \Delta\phi = \frac{\beta(s)\, \delta k(s)\, ds}{2}.$$

Since the betatron phase is not involved in this equation, we can just integrate around the ring to obtain

$$\Delta Q = \frac{1}{4\pi} \int \beta(s)\, \delta k(s)\, ds.$$

This is a very useful result and is worth memorizing. It is perhaps surprising that the change is independent of the phase of the perturbation. Strictly, this equation is only approximately true since, as we add each elemental focusing error, it modifies $\beta(s)$ as well as Q so that there is a higher-order term which should be included if one wants accurate numerical results (see Courant *et al.* 1958, Eqs (4.32)–(4.37)). Nevertheless, used with discretion, it is sufficiently accurate to explain the physical basis of the resonant phenomena discussed in later sections, since these can usually only be estimated to within a factor of 2.

6.4 Resonant conditions

The reason for our concern about the change in tune or phase advance which results from errors is that we must steer Q well away from certain fractional values which can cause the motion to resonate and result in loss of the beam. To understand how some Q values are dangerous, let us return to the case of closed-orbit distortion. Earlier we found the orbit distortion amplitude

$$\hat{y} = \frac{\sqrt{\beta\beta_{\mathrm{K}}}}{2\sin\pi Q} \cdot \frac{\Delta(Bl)}{B\rho}.$$

Clearly, this will become infinite if Q is an integer value. What happens physically is that the beam receives a kick at the same phase on every turn and just spirals outwards. An error in gradient can have the same effect if the Q value is close to one of the following lines:

$$2Q_\mathrm{h} = p, \qquad 2Q_\mathrm{v} = p,$$
$$Q_\mathrm{h} - Q_\mathrm{v} = p, \qquad Q_\mathrm{h} + Q_\mathrm{v} = p,$$

where p is an integer.

At this stage in the description of transverse dynamics we can only hint at the explanation for this. Particles spiral outwards in phase space if the perturbation has the same effect on each turn and this can only build up in this way if the particle returns to the same point in phase space on each turn ($Q = p$). The perturbation from a dipole is independent of the transverse displacement but a quadrupole error has field proportional to x and if a particle makes half-turns in phase space it will see alternately positive and negative kicks in divergence but both will reinforce the growth. This can happen if Q is a half-integer. One may extend this argument to understand why sextupole errors, which have a quadratic x dependence, excite the so-called third-integer 'resonances' near the following lines:

$$3Q_\mathrm{h} = p, \qquad 2Q_\mathrm{h} + Q_\mathrm{v} = p, \qquad Q_\mathrm{h} + 2Q_\mathrm{v} = p,$$
$$3Q_\mathrm{v} = p, \qquad 2Q_\mathrm{h} - Q_\mathrm{v} = p, \qquad Q_\mathrm{h} - 2Q_\mathrm{v} = p.$$

6.4.1 The working diagram

This is simply a diagram with Q_h and Q_v as its axes. The beam can be plotted on it as a point but, because there is a certain Q spread among particles of different momenta, we had better give the point a finite radius ΔQ (Fig. 6.12).

Figure 6.12 shows a mesh of lines which mark danger zones for the particles. We have hinted above that if Q in either the vertical or the horizontal plane is a simple fraction, then

$$nQ = p,$$

where n and p are integers and $n < 5$, a resonance takes over and walks the proton out of the beam. In general this is true when

$$lQ_\mathrm{h} + mQ_\mathrm{v} = p,$$

where $|l| + |m|$ is the order of the resonance and p is the azimuthal frequency which drives it. This equation just defines a set of lines in the Q diagram for each order of resonance and for each value of the integer p. Figure 6.12 shows these lines for the SPS.

Somehow, by careful adjustment of the quadrupoles in the lattice and by keeping the Q spread (chromaticity) small, we must coax the beam up to full energy without hitting the lines. To make things more difficult, each line has a

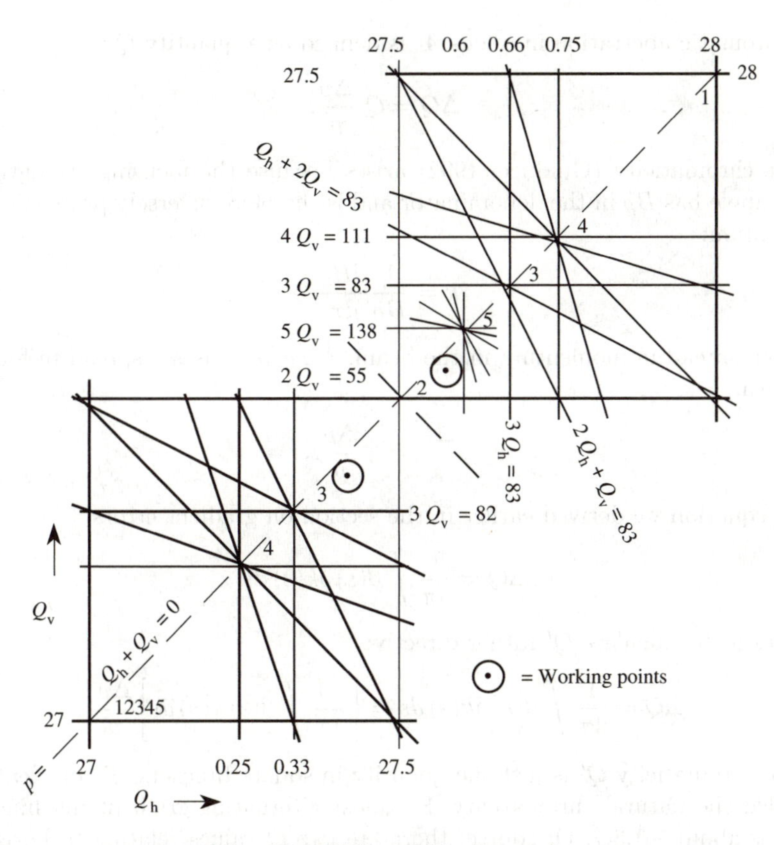

Fig. 6.12 SPS working diamond.

finite width, proportional to the strength of the imperfection which drives it. In some cases we must compensate the imperfections with correction multipoles to reduce this width.

We will discuss resonances and their correction in more detail in Chapter 7 but first a word about chromaticity.

6.5 Chromaticity

The steering of Q depends on careful regulation of quadrupole and dipole power supplies. In fact, much of the time for setting up a large circular accelerator is devoted to tuning Q to be constant as the fields and energy rise. Once beam has been accelerated the problem becomes one of reducing all effects which produce a spread in Q among the particles in the beam. The limit to this is usually reached when beam intensity is high enough to cause space-charge focusing effects, whose strength varies with the local beam density and which modulate Q as particles move up and down the bunch. Before reaching this limit, one must correct the tune spread due to momentum, the chromaticity. This is exactly equivalent to the chromatic aberration in a lens. It is defined as a quantity Q':

$$\Delta Q = Q' \frac{\Delta p}{p}.$$

The chromaticity (Guiducci 1992) arises because the focusing strength of a quadrupole has $B\rho$ in the denominator and is therefore inversely proportional to momentum:

$$k = \frac{1}{B\rho} \frac{dB_z}{dx}.$$

A small spread in momentum in the beam, $\pm\Delta p/p$, causes a spread in focusing strength:

$$\frac{\Delta k}{k} = \mp \frac{\Delta p}{p}.$$

An equation we derived earlier in the section on gradient errors

$$\Delta Q = \frac{1}{4\pi} \int \beta(s)\, \delta k(s)\, ds$$

enables us to calculate Q' rather directly:

$$\Delta Q = \frac{1}{4\pi} \int \beta(s)\, \delta k(s)\, ds = \left[\frac{-1}{4\pi} \int \beta(s)\, k(s)\, ds \right] \frac{\Delta p}{p}.$$

The chromaticity Q' is just the quantity in square brackets. To be clear, this is called the natural chromaticity. For most alternating-gradient machines, its value is about $-1.3Q$. Of course, there are two Q values relating to horizontal and vertical oscillations and therefore two chromaticities.

Chromaticity can be measured by changing the mean momentum of the beam by offsetting the r.f. frequency and measuring Q. Figure 6.13 shows such a measurement. One may calculate the link between r.f. frequency and momentum in a machine in which the bending field is held constant. Dispersion ensures that a higher momentum beam follows a larger mean radius and, above transition, has a lower revolution frequency. The r.f. frequency is an exact multiple of this higher revolution frequency. Conversely, if we change the r.f. frequency with the control system of the low-level cavity amplifier, the whole beam will be shifted to a different momentum at which the Q value will be different, thanks to chromaticity.

Imagine the situation at injection where $\Delta p/p$ can be $\pm 2 \times 10^{-3}$. In a synchrotron with a Q about 25, this can make the working point in the Q diagram into a line of length $\Delta Q = 0.15$. This is too long to avoid the resonances and must be corrected. In recent machines like LEP and LHC both Q and Q' are even higher.

One way to correct this is to introduce some focusing which gets stronger for the high-momentum orbits near the outside of the vacuum chamber—a quadrupole whose gradient increases with radial position—is needed.

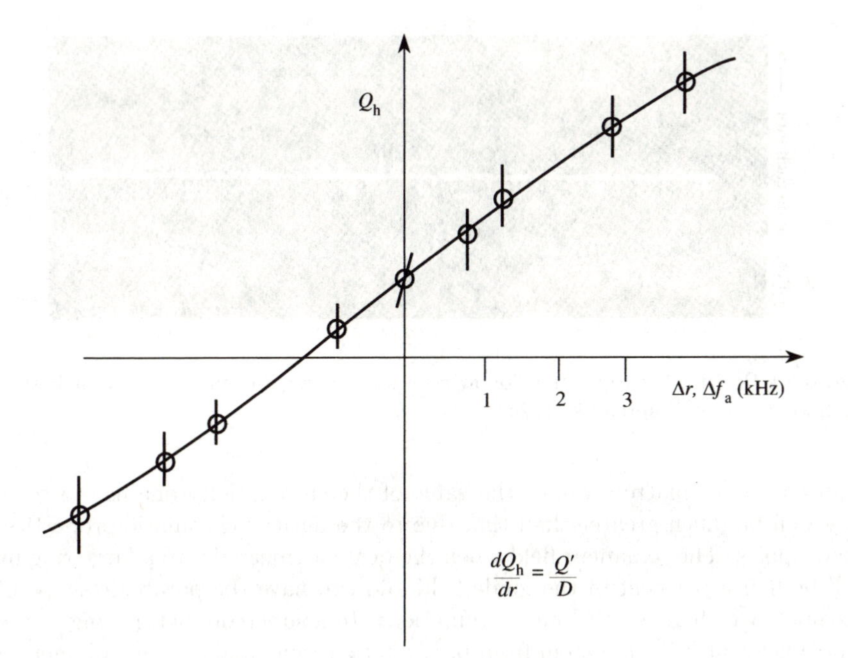

Fig. 6.13 Measurement of variation of Q made by changing the r.f. frequency with mean radius.

A sextupole whose field is

$$B_z = \frac{B''}{2}x^2$$

in a place where there is dispersion will introduce a normalized focusing correction:

$$\Delta k = \frac{B''D}{B\rho}\frac{\Delta p}{p}.$$

We use an earlier expression for the effect of this Δk on Q and obtain

$$\Delta Q = \left[\frac{1}{4\pi}\int\frac{B''(s)\beta(s)D(s)\,ds}{B\rho}\right]\frac{\Delta p}{p}.$$

To correct chromaticity we have to make the quantity in the square bracket balance the chromaticity. There are of course two chromaticities, one affecting Q_{h}, the other Q_{v} and we must therefore arrange for the sextupoles to cancel both. For this we use a trick which is common and will crop up again in other contexts. Sextupoles near F quadrupoles, where β_x is large, affect mainly the horizontal Q, while those near D quadrupoles, where β_z is large, influence Q_{v}. The effects of two families like this are not completely orthogonal but by inverting a simple 2×2 matrix one can find two sextupole sets which do the job.

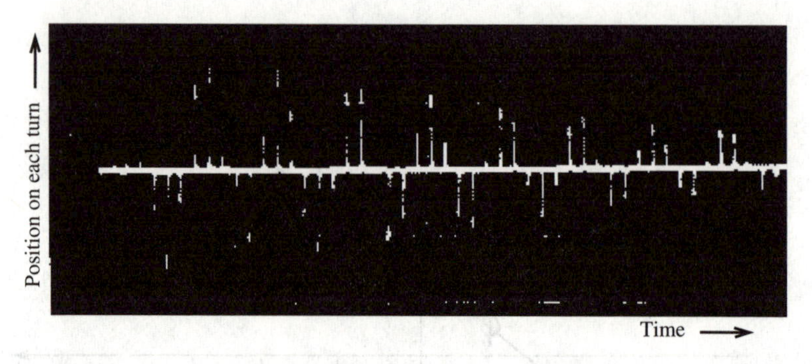

Fig. 6.14 Position pickup signal following a kick showing decay of coherent betatron oscillation due to Q spread $\approx 1/24$.

Just to make matters worse, the value of the chromaticity one has to correct may well be much greater than that due to the natural chromatic properties of quadrupoles. The remanent field when the beam is injected into a large ring may well be half a per cent of the guide field and can have the parabolic shape of a sextupole which generates pure chromaticity. In a superconducting ring the sextupole fields at injection stem from persistent currents and are much larger still.

Storage rings are usually designed with low-β sections with zero dispersion for the interaction regions and the main low-β quadrupoles, being very powerful, make strong contributions to the chromaticity. Since the dispersion is zero at the source of the error, the compensation can only be made elsewhere in the lattice where the parameter D is large.

The correction of chromaticity is a subject on its own and there is a higher-order term, a parabolic variation of Q with momentum, which is not compensated in this simple way. Sextupole patterns which minimize this, yet do not themselves excite serious non-linear side effects, are not easy to find.

There are two ways to measure chromaticity apart from the radial steering method shown in Fig. 6.13. The first of these is to observe the width of the betatron sidebands in the spectrum from a transverse pickup (Fig. 4.11). Secondly, we can measure the time it takes for a coherent betatron oscillation following a small kick to disappear as the ΔQ smears out the phase relation between protons of different momenta (Fig. 6.14). If the coherent oscillations persist for 200 turns, we may deduce that $\Delta Q \approx 1/200$ and is about the best we can hope for using this rather crude method. In the chapter on instabilities, we shall return to the concept of the decay of a coherent signal for it is the basis of Landau damping.

Exercises

6.1 Given that the SPS vacuum chamber half-height is 28 mm at a point where the lattice $\beta = 104$ m: What emittance will just touch it? How big must the vacuum chamber be at a vertically defocusing quadrupole where $\beta = 20$ m? What is the maximum vertical divergence of the beam which just touches the chamber at $\beta = 104\,\text{m}$? Given that the magnetic rigidity is 1334 T m at 400 GeV/c, what magnetic field is needed in a 20 m long dipole designed to deflect the whole beam into the wall of the vacuum chamber. (Hint: Deflect the phase ellipse by its full spread in divergence.)

6.2 Two dipoles of equal strength are located at F quadrupoles 180° apart in ϕ in the SPS. Their purpose is to deflect the beam into the vacuum chamber as they are slowly switched on. Draw a circle diagram to indicate the trajectory of a particle as it passes between these magnets.

6.3 The tolerance on quadrupole misalignment in the SPS is 0.15 mm and the Q is nominally 27.75. What amplitude of distortion will be produced for $B\rho = 1334$ (T m) at a β of 108 m if: (a) One quadrupole, of length 3.22 m and normalized strength, $k = 20\,(\text{T m})/(B\rho)$ is displaced by 0.15 mm at a β of 108 m. (b) There are 108 such quadrupoles with an r.m.s. displacement of 0.15 mm. How big do you expect the closed-orbit distortion to be before correction (allow a factor of 2 for safety)? (c) How big would this be if you also include the 108 D quadrupoles, where $\beta = 20\,\text{m}$? (d) What r.m.s. tilt of the 744 dipoles is comparable to the quadrupoles' effect (assume a mean β value at the dipoles of 64 m)?

6.4 The working point chosen for a proton storage ring is 27.61, 27.63. Discuss the nearby resonance lines. If the natural chromaticity is 36 and uncorrected how large a momentum spread would avoid touching the nearest lines.

6.5 Take the parameters of the synchrotron in Exercise 6.1 and assume there are six equispaced sextupoles to correct a chromaticity of 10 units at injection. They are placed at positions where $D = 1.2\,\text{m}$, $\beta = 109\,\text{m}$. Each sextupole is 0.3 m long. What would the sextupole gradient B'' be?

6.6 If the maximum field on the pole tip (at radius, a) is 0.8 T and the sextupole aperture bore $= 2 \times 0.08\,\text{m}$, how high a $B\rho$ or momentum will they correct?

6.7 If we install a second set to correct $Q'_{\text{v}} = 36$ at D quads where $\beta_{\text{v}} = 109$ but $D = 0.2\,\text{m}$, how much smaller aperture must they have?

7 Non-linearities and resonances

7.1 Introduction

In our study of transverse dynamics so far, we have ignored non-linearities in the guide field and assumed that the bending magnets have a pure dipole field and gradient magnets or quadrupoles have a radial field which has a constant linear slope. Non-linear fields due to higher-order multipole terms have not been considered. However, if we were fixing the parameters of a real machine, we could not go too far without considering the practical difficulties of designing the magnets and determining the tolerances which can be reasonably written into the engineering specification (Blewett 1977; Wilson 1984). Estimates of the non-linear departures from pure dipole or gradient field shape, and of the statistical fluctuation of these errors around the ring at each field level must be made. We must take into consideration that the remnant field of a magnet may have quite a different shape from that defined by the pole geometry; the steel properties may vary during the production of laminations; the eddy currents in vacuum chamber and coils may perturb the linear field shape. Mechanical tolerances must be chosen to ensure that asymmetries do not creep in. At high field, the linearity may deteriorate owing to saturation, and variations in packing factor of laminations can become important. Superconducting magnets will have strong error fields due to persistent currents in their coils whose effect can be serious at injection and high-order multipoles due to conductor must be under control.

When these effects have been reviewed, tolerances and assembly errors may have to be revised and measures taken to mix or match batches of laminations with different steel properties or coils made from different batches of superconductor. It may be necessary to place magnets in a particular order in the ring in the light of production measurements of field uniformity or to shim some magnets at the edge of the statistical distribution. Even when all these precautions have been taken, non-linear errors may remain whose effect will have to be compensated with auxiliary multipole magnets. We must have good estimates of the effects of these errors to design the correcting magnets.

The principal reason for all this is not only to minimize closed-orbit distortion but to reduce the influence of non-linear resonances on the beam. A glance at the working diagram shows why this is so. In Fig. 6.12 we saw that the working diagram is traversed by a mesh of non-linear resonance lines or stopbands of first, second, third, and fourth order. Each resonance line is driven by a particular pattern of multipole field errors which can be present in the guide field. The order, n, determines the spacing in the Q diagram; third-order stopbands, for instance, converge on a point which occurs at every $1/3$ integer Q value (including the integer itself). The order n is also related to the order of the non-linearity or multipole which drives the resonance. For example, fourth-order resonances are driven by multipoles with $2n$ poles, that is, octupoles.

The strength of the non-linear resonances driven by non-linear multipoles is amplitude dependent so they become more important as we seek to use more and more of the machine aperture. We can think of them as defining a 'dynamic' aperture which, in general, is smaller than the beam tube. Theorists used to discount resonances of fifth and higher order as harmless (self-stabilized), but experience in large hadron storage rings indicates that this may not be a good assumption when we want beams to circulate for more than a few seconds.

The resonance lines have a finite width which depends directly on the strength of the error. This width increases with amplitude. We must ensure that the errors are small enough to leave some clear space between the stopbands to tune the machine, otherwise particles will fall into the stopbands and will be ejected before they have, even, been accelerated. The stopband widths are mainly influenced by the random fluctuations in multipole error around the ring rather than the mean multipole strength.

However, systematic or average non-linear field errors can also make life difficult. They cause Q to be different for the different particles in the beam, depending upon their betatron amplitude or momentum defect. Such a Q spread implies that the beam will need an even larger resonance-free window in the Q diagram. In the case of the large machines, SPS, HERA, LEP, and LHC, the window would not be large enough if we did not balance out the average multipole component in the ring by powering multipole correction magnets in addition to the sextupoles which control chromaticity.

Paradoxically, when a 'pure' machine has been designed and built, there is often a need to impose a controlled amount of non-linearity to correct the momentum dependence of Q or to introduce a Q spread among the protons to prevent a high-intensity instability.

7.2 Multipole fields

Before we come to discuss the non-linear terms in the dynamics, we need to describe the field errors which drive them. The magnetic *vector potential* of a magnet with $2n$ poles in Cartesian coordinates is

$$\mathbf{A} = \sum_n A_n f_n(x, z),$$

Table 7.1 Cartesian solutions of magnetic vector potential

Multipole	n	Regular f_n	Skew f_n
Quadrupole	2	$x^2 - z^2$	$2xz$
Sextupole	3	$x^3 - 3xz^2$	$3x^2z - z^3$
Octupole	4	$x^4 - 6x^2z^2 + z^4$	$4x^3z - 4xz^3$
Decapole	5	$x^5 - 10x^3z^2 + 5xz^4$	$5x^4z - 10x^2z^3 + z^5$

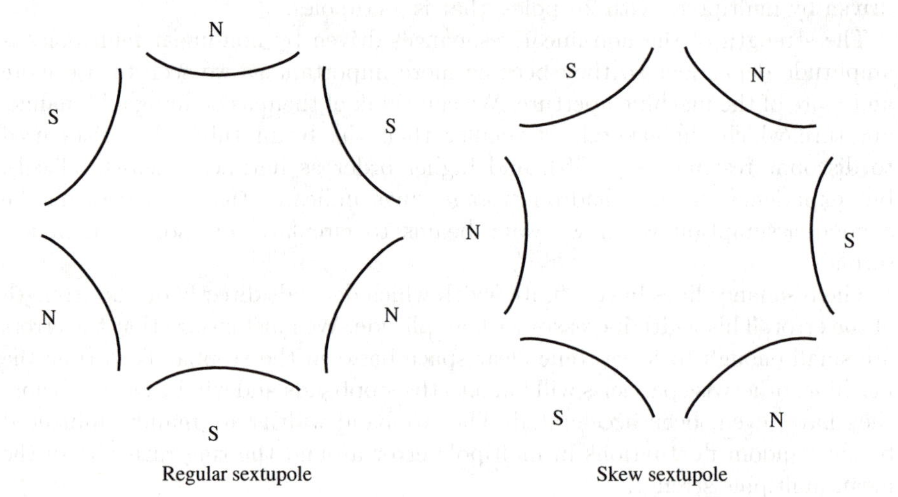

Fig. 7.1 Pole configurations for a regular sextupole and a skew sextupole.

where f_n is a homogeneous function in x and z of order n. If the magnet is long we can ignore end fields and assume that the vector potential is entirely parallel to the magnet axis A_s.

Table 7.1 gives $f_n(x, z)$ for both regular and skew, low-order multipoles. Figure 7.1 shows the distinction between regular and skew multipoles. We can obtain the function for other multipoles from the binomial expansion of

$$f_n(x, z) = (x + iz)^n.$$

The real terms correspond to regular multipoles, the imaginary ones to skew multipoles.

Of course, we are interested in the field and, for a long regular magnet with a purely paraxial potential,

$$B_z(z = 0) = -\frac{\partial A_\mathrm{s}}{\partial x} = -\sum_{n=1}^{\infty} n A_n x^{n-1}$$

$$= -\sum_{n=1}^{\infty} \frac{1}{(n-1)!} \left(\frac{d^{n-1} B_z}{dx^{n-1}} \right)_0 x^{n-1}.$$

In the last line we have used a Taylor expansion to express the field, and, by equating these expressions we find

$$A_n = \frac{1}{n!}\left(\frac{d^{n-1}B_z}{dx^{n-1}}\right)_0.$$

A more modern convention (in Europe) is to speak of multipole coefficients b_n for normal components and a_n for skew components, where R_r is some reference radius (10 mm for the LHC), B_1 is the magnitude of the nominal dipole field B_y, and $Z = x + iz$:

$$B_y + iB_x = B_1 \sum_n (b_n + ia_n)\left(\frac{Z}{R_\mathrm{r}}\right)^{n-1}.$$

The suffix $n = 1$ for the dipole, 2 for the quadrupole, and 3 for sextupole, etc. Note that the US notation starts with $n = 0$ for the dipole and different laboratories use other reference radii. In spite of the possibilities for confusion, this has the advantage that the coefficients are a measure of the tolerated fraction of field error inside the reference circle where the beam is supposed to be stable. It is, therefore, much easier to compare the designs of the different machines.

7.2.1 Field symmetry

We digress a little to discuss the sextupole errors in the main dipoles of a large synchrotron as a practical example of how one may identify the multipole components of a magnet by inspecting its symmetry. Let us consider a simple dipole (Fig. 7.2). It is symmetric about the vertical axis and its field distribution contains mainly even exponents of x, corresponding to odd n values in the vector potential: dipole, sextupole, decapole, etc. We also see that cutting off the poles symmetrically at a finite width can produce a virtual sextupole. Moreover, the remanent field pattern is frozen in at high field where the flux lines leading to the pole edges are shorter than those leading to the centre. The remanent magnetomotive force,

$$\int H_\mathrm{c}\, dl,$$

is weaker at the pole edges, and the field in the gap tends to sag into a parabolic configuration. This too produces a sextupole.

Very similar sextupole effects are generated by the persistent currents in the windings of a superconducting magnet. These eddy currents in the coils depress the edges of the pure dipole field to produce strong six- and ten-pole effects.

Such sources of sextupole error are the principal non-linearities in a large machine. Note that they have no skew component. However, before considering the non-linearities further, let us examine a simple resonance which is strictly linear.

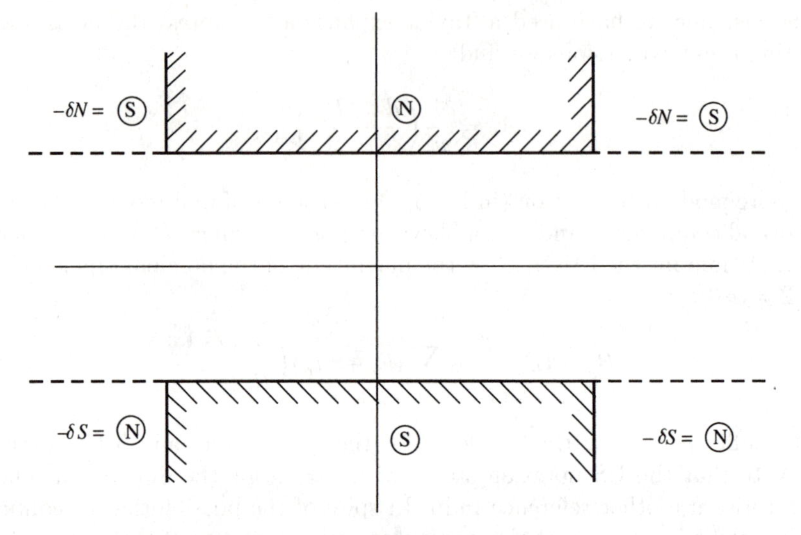

Fig. 7.2 The field in a simple dipole. The δN and δS poles superimposed on the magnet poles give the effect of cutting off the poles to a finite width.

7.3 The second-order resonance

We shall use the circle approximation to solve the problem of the resonance driven by quadrupoles. A small elementary quadrupole of strength $\delta(Kl)$ is located close to an F quadrupole where $\beta_\mathrm{h} = \hat{\beta}$ (Fig. 7.3). Suppose a particle describes a circular trajectory of radius $a = \sqrt{\varepsilon\beta}$ and encounters the quadrupole at phase

$$\varphi(s) = Q\theta,$$

where θ is the azimuth which corresponds exactly to φ at the quadrupoles of a FODO lattice.

The first step is to write down the unperturbed displacement at the small quadrupole:

$$x = a\cos Q\theta.$$

The particle receives a divergence kick (Fig. 7.3):

$$\Delta x' = \frac{\Delta(Bl)}{B\rho} = \frac{\Delta(Kl)x}{B\rho}.$$

The small change in $\hat{\beta}\Delta x'$,

$$\Delta p = \hat{\beta}\Delta x',$$

perturbs the amplitude a by

$$\Delta a = \Delta p \sin Q\theta.$$

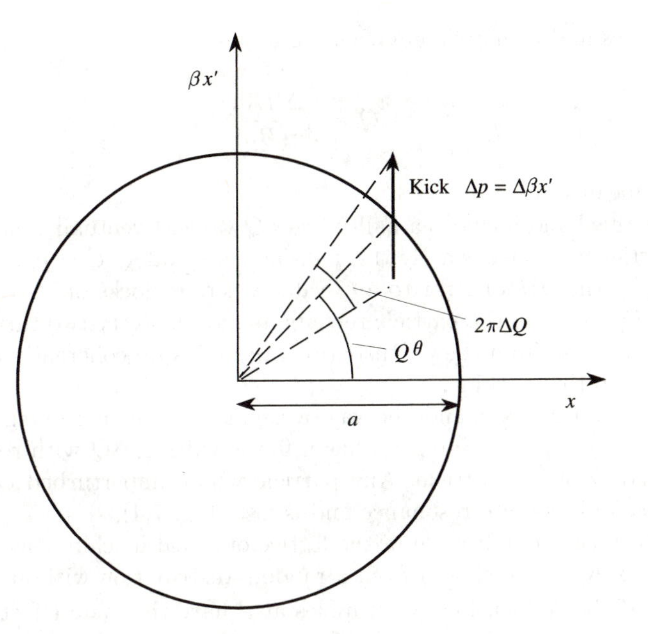

Fig. 7.3 Circle diagram shows effect of kick at phase $Q\theta$. There is a small phase advance.

Even more significantly, there is a small phase advance (Fig. 7.3):

$$2\pi\Delta Q = \frac{\Delta p}{a}\cos Q\theta.$$

By successive substitution, we get

$$2\pi\Delta Q = \hat{\beta}\frac{\Delta(lK)}{B\rho}\cos^2 Q\theta.$$

Over one turn, the Q changes from the unperturbed Q by

$$\Delta Q = \frac{\hat{\beta}\Delta(lK)}{4\pi(B\rho)}(\cos 2Q\theta + 1).$$

On an average this shifts Q by

$$\Delta Q = \frac{\hat{\beta}\Delta(lK)}{4\pi(B\rho)}.$$

Similarly, the average perturbation in amplitude a is

$$\frac{\Delta a}{a} \approx 2\pi\Delta Q.$$

However, the phase $Q\theta$ on which the particle meets the quadrupole changes on each turn by 2π times the fractional part of Q and hence the Q value for each turn oscillates and it may lie anywhere in a band

$$\delta Q = \frac{\hat{\beta}\Delta(lK)}{4\pi(B\rho)}$$

about the mean value.

Suppose this band includes a half-integer Q value. Eventually, on a particular turn, a particle will have exactly this half-integer Q value ($Q = p/2$). It has been perturbed by the $\Delta(lK)$ error to a Q value where it 'locks on' to a half-integer stopband. Once there, the particle repeats its motion every two turns, the small amplitude increase from the perturbation Δa builds up coherently and extracts the beam from the machine.

We can visualize this in another way by saying that the half-integer line in the Q diagram, $2Q = p$ (p = integer), has a finite width $\pm\Delta Q$ with respect to the unperturbed Q of the particle. Any particle whose unperturbed Q lies in this *stopband width* locks into resonance and is lost (Fig. 7.4).

In practice, each quadrupole in the lattice of a real machine has a small field error. The $\Delta(Kl)$'s are chosen from a random distribution with an r.m.s. value $\Delta(Kl)_{\mathrm{rms}}$. If the N focusing quadrupoles at $\hat{\beta}$ have the main effect, we can see that the r.m.s. expectation value for δQ is

$$\langle\delta Q\rangle_{\mathrm{rms}} = \sqrt{\frac{N}{2}}\frac{\hat{\beta}\Delta(Kl)_{\mathrm{rms}}}{4\pi(B\rho)}.$$

The factor of $\sqrt{2}$ comes from integrating over the random phase distribution. The statistical treatment is similar to that used for estimating the closed-orbit distortion.

Now let us use Fourier analysis to see which particular azimuthal harmonic of the $\Delta(Kl)$ pattern drives the stopband. Using the normalized strength $k = K/(B\rho)$, we analyse the function $\delta(\beta k)$ into its Fourier harmonics

$$\delta\beta k(s) = \sum \hat{\beta}k_p \cos(p\theta + \lambda),$$

and

$$\hat{\beta}k_p = \frac{1}{\pi R}\int_0^{2\pi} ds\, \delta[\beta k(s)] \cos(p\theta + \lambda).$$

In general, all harmonics, that is, all values of p, have equal expectation values in the random pattern of errors. We substitute the pth term in the series into the earlier expression for the increment in Q and work through the steps to obtain

$$2\pi\Delta Q = \int \frac{\hat{\beta}k_p}{2} \cos(p\theta + \lambda)\{\cos 2Q\theta + 1\}\, ds.$$

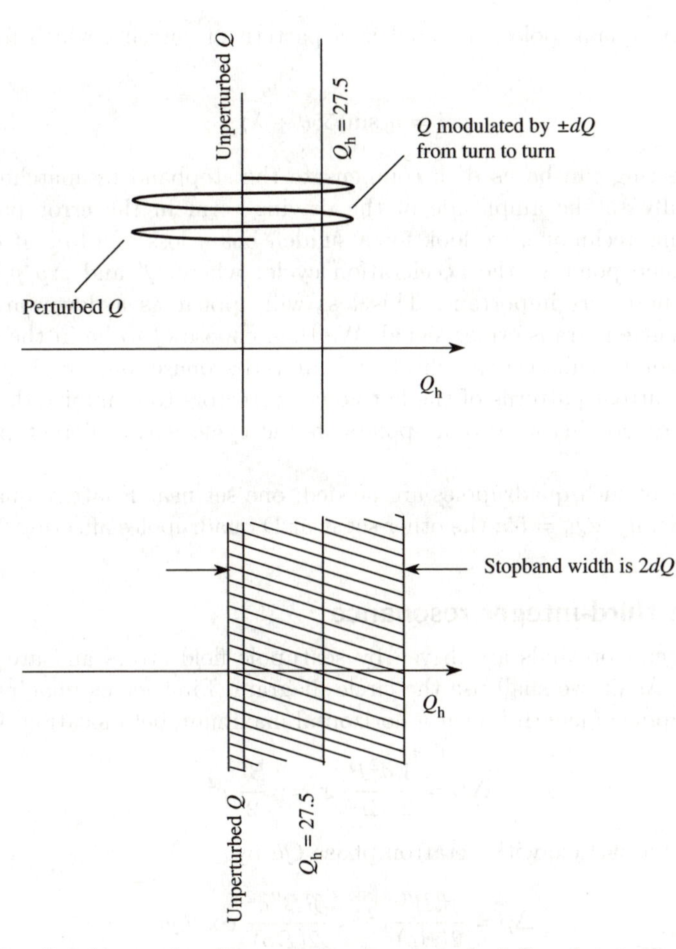

Fig. 7.4 Alternative diagrams showing perturbed Q and a stopband.

The integration can be simplified by writing $ds = R\,d\theta$:

$$\Delta Q = \frac{\hat{\beta} k_p R}{4\pi} \int_0^{2\pi} \cos 2Q\theta \, \cos(p\theta + \lambda)\, d\theta.$$

The integral is finite only over many betatron oscillations when the resonant condition is fulfilled:

$$2Q = p.$$

We have revealed the link between the azimuthal frequency p in the pattern of quadrupole errors and the $2Q = p$ condition which describes the stopband. For example, in the SPS the half-integer stopband $2Q = 55$ lies close to $Q = 27.6$. The azimuthal Fourier component which drives this is $p = 55$. Similarly, a pattern of correction quadrupoles, powered in a pattern of currents which follows the function

$$i = i_0 \sin(55\theta + \lambda)$$

around the ring can be used to compensate the stopband by matching i_0 and λ empirically to the amplitude of the driving term in the error pattern. In applying this technique, we look for a sudden beam loss due to a strong stopband at some point in the acceleration cycle, where Q' and $\Delta p/p$ are large and gradient errors important. This loss will appear as a downward step in the beam current transformer signal. We then choose Q to lie in the stopband at that point to enhance the effect and alter the phase and amplitude of the azimuthal current patterns of the harmonic correctors to minimize the loss. We may have to do this at various points in the cycle with different phase and amplitude.

Two sets of such quadrupoles are needed: one set near F lattice quadrupoles affecting mainly $2Q_{\mathrm{h}} = 55$; the other set near D quadrupoles affecting $2Q_{\mathrm{v}} = 55$.

7.4 The third-integer resonance

Third-integer stopbands are driven by sextupole field errors and are therefore non-linear. Again we shall use the circle diagram. First let us imagine a single short sextupole of length l, near a horizontal maximum beta location. Its field is

$$\Delta B = \frac{1}{2}\frac{d^2 B_z}{dx^2}x^2 = \frac{B''}{2}x^2,$$

and it kicks a particle with betatron phase $Q\theta$ by

$$\Delta p = \frac{\beta l B''}{2(B\rho)}x^2 = \frac{\beta l B'' a^2}{2(B\rho)}\cos^2 Q\theta$$

inducing the following increments in phase and amplitude:

$$\frac{\Delta a}{a} = \frac{\Delta p}{a}\sin Q\theta = \frac{\beta l B'' a}{2(B\rho)}\cos^2 Q\theta \sin Q\theta,$$

$$\Delta\phi = \frac{\Delta p}{a}\cos Q\theta = \frac{\beta l B'' a}{2(B\rho)}\cos^3 Q\theta$$

$$= \frac{\beta l B'' a}{8(B\rho)}(\cos 3Q\theta + 3\cos Q\theta).$$

Suppose Q is close to a third integer; then the kicks on three successive turns appear as in Fig. 7.5. The second term averages to zero over three turns and we are left with a phase shift

$$2\pi\Delta Q = \Delta\phi = \frac{\beta l B'' a \cos 3Q\theta}{8(B\rho)}.$$

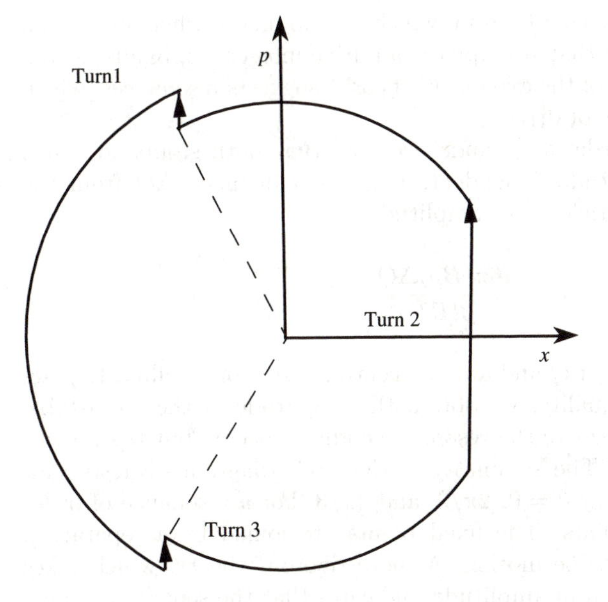

Fig. 7.5 Phase-space trajectory on a third-order resonance.

We can now guess how resonances arise. Close to $Q = p/3$, where p is an integer, $\cos 3Q\theta$ varies slowly, wandering within a band about the unperturbed Q_0 as in Fig. 7.4:

$$Q_0 - \frac{\beta l B'' a}{16\pi(B\rho)} < Q < Q_0 + \frac{\beta l B'' a}{16\pi(B\rho)}.$$

As in the case of the half-integer resonance we call this the stopband width but it really is a perturbation in the motion of the particle itself. We can write the expression for amplitude perturbation

$$\frac{\Delta a}{a} = \frac{\beta l B'' a}{8(B\rho)} \sin 3Q\theta.$$

Suppose the third-integer Q value is somewhere in the band. Then, after a sufficient number of turns, the perturbed Q of the machine will be modulated to coincide with $Q \approx p/3$. On each subsequent revolution, this increment builds up until the particle is lost. Growth is rapid and the modulation of Q away from the resonant line is comparatively slow.

Looking back at the expressions earlier in this analysis, we find that the resonant condition, $3Q$ = integer, arises because of the $\cos^3 Q\theta$ term, which in turn stems from the x^2 dependence of the sextupole field. This reveals the link between the order of the multipole and that of the resonance. We see that the a^2 leads to a linear dependence of the width on the amplitude. The equivalent term was a^1 in the case of the half-integer resonance, which led to a width that was independent of amplitude. The same term will become a^3 in the case of a fourth-order resonance giving a parabolic dependence of width upon amplitude.

It is worth noting that the $\cos Q\theta$ term, which we can ignore when away from an integer Q value, suggests that sextupoles can drive integer stopbands as well as third integers. Inspection of the expansion of $\cos^n\theta$ suggests resonances, which other multipoles are capable of driving.

Returning to the third-order stopbands, we note that both stopband width and growth rate are amplitude dependent. If Q_0 is a distance ΔQ from the third-integer resonance, particles with amplitude

$$a < \frac{16\pi(B\rho)\Delta Q}{\beta l B''}$$

never reach a one-third integer Q and are in a central region of stability. Replacing the inequality by an equality, we obtain the amplitude of the metastable fixed points in phase space where the resonant condition occurs but the growth is infinitely slow (Fig. 7.6). The symmetry of the circle diagram suggests that there are three fixed points at $\theta = 0$, $2\pi/3$, and $4\pi/3$. For a resonance of order n, there will be n such points. The fixed points are joined by a separatrix, which is the boundary of stable motion. A more rigorous theory, which takes into account the perturbation in amplitude, indicates that the separatrix shape is triangular, with the three arms being those to which particles cling on their way out of the machine.

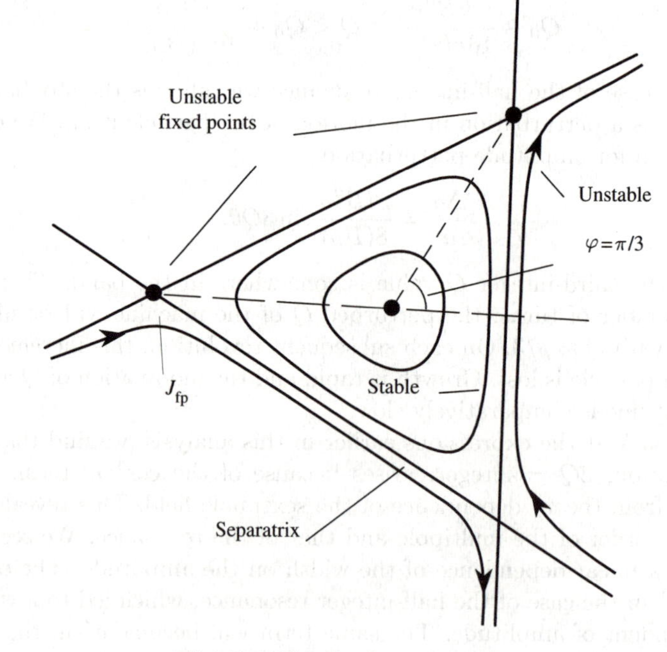

Fig. 7.6 Third-order separatrix.

We have seen how a single sextupole can drive the resonance. Suppose now we have an azimuthal distribution of sextupoles which can be expressed as a Fourier series:

$$B''(\theta) = \sum B''_p \cos p\theta,$$

then

$$\Delta\phi = \sum_p \int \frac{\beta B''_p}{8(B\rho)} \cos 3Q\theta \, \cos p\theta \, d\theta.$$

This integral is large and finite if $p = 3Q$.

This reveals, as in the earlier case of the second-order resonances, why it is a particular harmonic in the azimuthal distribution which drives the stopband. Strictly it is not the Fourier spectrum of $B''(\theta)$, but of $\beta B''(\theta)$ which is important. Periodicities in the lattice reflected in the shape of $\beta(\theta)$ and the periodicity of the multipole pattern can thus mix to drive the resonances. This is particularly important since some multipole fields, like the remanent field pattern of dipole magnets, are inevitably distributed in a systematic pattern around the ring. This pattern is rich in the harmonics of S, the superperiodicity. Even if the errors are evenly distributed, any modulation of β which follows the pattern of insertions can give rise to systematic driving terms. It is an excellent working rule to keep any systematic resonance, that is,

$$3Q = S(\text{superperiod number}) \times \text{integer} = p,$$

out of the half-integer square in which Q is situated. This is often not easy in practice.

If we were prepared to wade through a more lengthy analysis (Guignard 1970; 1978), we would find that the resonant condition $3Q = p$ can be generalized into $nQ = p$, where $2n$ is the number of poles in the driving field pattern. Moreover, both Q_h and Q_v are involved and each multipole can drive $2n$ lines in the working diagram corresponding to each value of p. These are

$$lQ_\mathrm{h} + mQ_\mathrm{v} = p,$$

where $|l| + |m| = n$. In the third-integer case these are

$$3Q_\mathrm{h}, \quad 2Q_\mathrm{h} \pm Q_\mathrm{v}, \quad Q_\mathrm{h} \pm 2Q_\mathrm{v}, \quad 3Q_\mathrm{v} = p.$$

The positive sign defines 'sum' resonances and the negative sign the 'difference' resonances. These alternate between those driven by normal and skew multipoles (Fig. 7.1).

As in the second-order case, the four third-order sum resonances can be compensated with sets of multipoles powered individually to generate a particular Fourier component in their azimuthal distributions. By permuting these two kinds of sextupoles with the two types of location, we can attack the four lines more or less orthogonally.

7.4.1 Slow extraction using the third-order resonance

So far we have thought of resonances as a disease to be avoided, yet there is at least one useful function that they can perform. Slow extraction, though rarely needed in large synchrotron colliders, is used to deliver a steady beam from medical synchrotrons.

We have seen that a third-order stopband extracts particles above a certain amplitude, the amplitude of the unstable fixed points which define a separatrix between stability and instability (Fig. 7.6). The dimensions of the separatrix characterized by a are determined by ΔQ, the difference between the unperturbed Q and the stopband. As one approaches the third integer by, say, increasing the focusing strength of the lattice quadrupoles, ΔQ shrinks, the unstable amplitude a becomes smaller, and the particles are squeezed out along the three arms of the separatrix. If we make ΔQ shrink to zero over a period of a few hundred milliseconds, we can produce a rather slow spill extraction.

At first sight, we might expect only one-third of the particles to migrate to positive x values since there are three separatrices, but we must remember that a particle jumps from one arm to the next each turn, finally jumping the extraction septum on the turn when its displacement is largest. The septum is a thin-walled deflecting magnet at the edge of the aperture. The growth increases rapidly as particles progress along the unstable separatrix and, if the stable area is small compared to the distance between the beam and the septum, the growth per turn will be large and the probability of a particle striking the septum rather than jumping over it is small. Clearly, it also helps to have a thin septum.

The magnet or quadrupole ripple can cause an uneven spill, making Q approach the third integer in a series of jerks and modulating the rate at which particles emerge. However, a spread in particle momentum will smooth this out and, if the chromaticity is finite, we will have swept through a much larger range of Q values before all separatrices for all momenta have shrunk to zero. The change in Q caused by ripple is then much less significant.

7.5 Landau damping with octupoles

Another beneficial effect of multipoles is the use of octupoles to damp the coherent transverse collective instabilities.

We shall later see in Chapter 9 that for a transverse instability to be dangerous, the growth time must win over other mechanisms which tend to destroy the coherent pattern and damp out the motion. One such damping mechanism is the Q spread in the beam. Coherent oscillations decay, or become dephased, in a number of betatron oscillations comparable to $1/\Delta Q$, where ΔQ is the Q spread in the beam. This corresponds to a damping time, expressed in terms of the revolution frequency, $\omega_0/2\pi$:

$$\tau_{\mathrm{d}} = \frac{2\pi}{\omega_0 \Delta Q},$$

which is just the inverse of the spread in frequencies of the oscillating particles. The threshold for the growth of the instability is exceeded when its time τ_{g}, which increases with intensity, becomes faster than τ_{d}. This is an example of *Landau damping.*

Unfortunately, as a result of improvements in single-particle dynamics and as we correct chromaticity in our quest for a small ΔQ to avoid lines in the Q diagram, we inevitably lower the threshold intensity for the instability. A pure machine is infinitely unstable. Typically, as the optics is refined, suddenly the beam begins to suffer from an instability and a large fraction is lost before stability is restored at a lower intensity. Octupoles, however, produce an amplitude Q-dependence which is more effective than the momentum-dependent Q spread produced by sextupoles in damping instabilities. The sextupoles are less effective because each particle changes in momentum during a synchrotron oscillation and, in a time comparable to τ_{g}, all particles have the same mean momentum. Sextupoles do not therefore spread the mean Q of the particles. Octupoles, producing an amplitude Q-dependence, do.

The circle diagram can be used to calculate the effect of an octupole which gives a kick

$$\Delta p = \beta \frac{\Delta(Bl)}{B\rho} = \frac{\beta l B'''}{3! B\rho} a^3 \cos^3 Q\theta.$$

The change in phase is

$$2\pi\Delta Q = \Delta\phi = \frac{\beta l B''' a^2 \cos^4 Q\theta}{6(B\rho)},$$

which averages to

$$\Delta Q = \frac{\beta l B''' a^2}{32\pi(B\rho)}.$$

Of course, if the octupoles are placed around the ring, they can also excite fourth-order resonances. A good rule is to have as many of them as possible and to distribute them at equal intervals of betatron phase. If there are S octupoles distributed in this way, their Fourier harmonics are S, $2S$, etc. and they can only excite structure resonances near the Q values given by

$$4Q = S \times \text{an integer}.$$

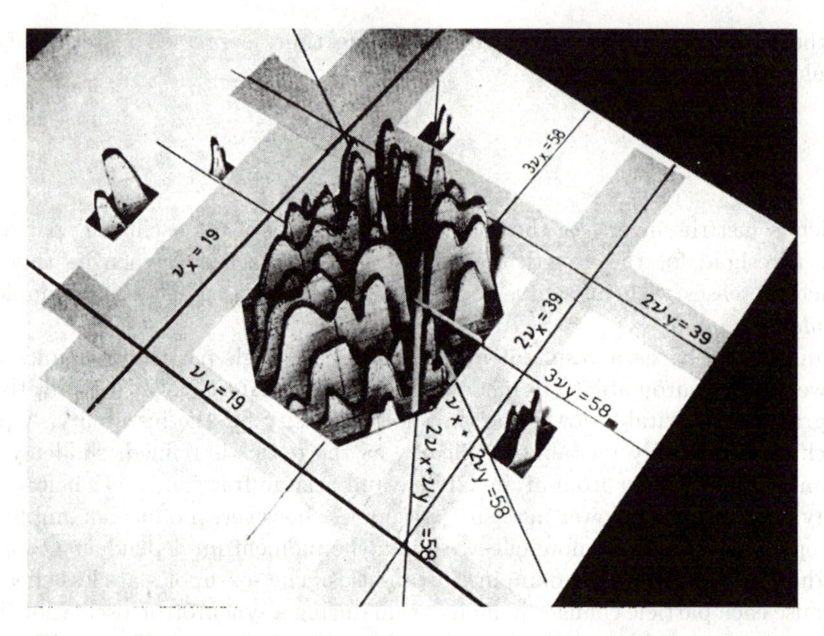

Fig. 7.7 Beam survival impaired by resonances in the FNAL main ring.

Although these systematic resonances are very strong, it should not be difficult to choose S so that Q is not in the same integer square as one of the values of $nS/4$.

7.6 Injection studies at FNAL

As a cautionary tale to complete this analysis of resonant behaviour, we should examine the contour model of beam survival in Fig. 7.7. This is the result of an old but important experiment performed when the FNAL main ring was first commissioned. It shows how uncorrected resonances can severely limit the beam which is accelerated. The third-integer lines (ν, the US nomenclature is equivalent to European Q) reduce beam survival to zero in some regions.

In this case, magnet ripple, another effect which scales adversely with ring size, ensured that the resonances were broadened making it virtually impossible to obtain full survival. It should be said that at the time this machine was built, accelerator experts had been lulled into complacency about resonances by experiences with smaller rings. Fortunately, it was possible to compensate the resonances and the FNAL machine went on to break all intensity records.

Exercises

7.1 A synchrotron consists of 108 HF (horizontally focusing) quadrupoles ($k > 1$) alternating with 108 HD quadrupoles. The gradient of the quadrupoles is 19.4 T/m at 400 GeV/c momentum. They are 3.22 m long. The maximum horizontal β is at HF and is 108 m, and the minimum β is at HD and is 18 m. In the vertical plane the HF become defocusing and HD become focusing. Hence, the same vertical maximum in β is at HD and the same minimum at HF. Devise a 2×2 matrix which give the fractional current change, in the separate circuits which feed HF and HD quadrupoles, necessary for a small change in Q_{h} and Q_{v}.

7.2 What is the chromaticity, $dQ/d(\Delta p/p)$, of this machine?

7.3 A sextupole field can be expressed as

$$B_y = \frac{B''}{2}x^2.$$

Derive an expression for the local quadrupole field about a particle whose displacement is x in the horizontal plane.

7.4 Suppose there are 36 such magnets, 85 cm long with $B'' = 150\,\mathrm{T/m^2}$, placed next to HF quadrupoles at a mean dispersion of 2.2 m. How much will this change the horizontal Q for a 400 GeV/c particle with momentum error of 1‰?

7.5 How much is the change in horizontal chromaticity introduced by these sextupoles?

7.6 What happens to the vertical chromaticity?

7.7 How can you simultaneously correct

$$\frac{dQ_{\mathrm{h}}}{d(\Delta p/p)} \quad \text{and} \quad \frac{dQ_{\mathrm{v}}}{d(\Delta p/p)}?$$

7.8 What is the 'n' value for the magnet shown below?

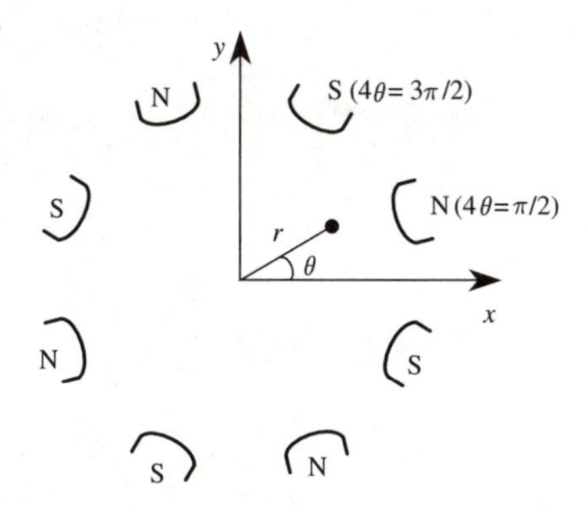

7.9 Write down an expression for the magnetic scalar potential, ϕ, in polar co-ordinates for this magnet given that

$$B_x = \frac{\partial \phi}{\partial x}, \qquad B_y = \frac{\partial \phi}{\partial y}$$

and that

$$\frac{\partial \phi}{\partial x} = \cos\theta \frac{\partial \phi}{\partial r} - \frac{\sin\theta}{r}\frac{\partial \phi}{\partial \theta},$$
$$\frac{\partial \phi}{\partial y} = \sin\theta \frac{\partial \phi}{\partial r} + \frac{\cos\theta}{r}\frac{\partial \phi}{\partial \theta}.$$

Find simple expressions for B_x and B_y as a function of x and y. How would you write ϕ as a function of x and y?

7.10 The magnet in exercise 7.8 has a horizontal variation of B_y which is a simple function of x. What is that function? Use a circle diagram to find an expression for the stopband width and Q shift from an octupole remembering that

$$8\cos^4\theta = \cos 4\theta + 4\cos\theta + 3.$$

What is the essential difference in the physics of the behaviour of a particle in an octupole in comparison with a quadrupole?

8 Electrons

8.1 Synchrotron radiation

One of the first decisions to be made in designing a synchrotron is to choose the maximum field and its radius. We have seen how the product of these two quantities, $B\rho$, is proportional to the momentum of the particle and this, in turn, to the energy of the particle, if it is close to the velocity of light. During the history of the development of synchrotrons, these two limits have restricted designers to a window of energy from roughly 1 GeV to a few TeV.

So far in this study we have not had to consider whether the synchrotron is to accelerate electrons or protons. Indeed nearly all the theory which we have presented applies equally to both kinds of particles. There is, however, one important difference which we must examine in the context of synchrotron radiation. The mass of the electron is roughly 2000 times smaller than that of the proton and, therefore, for the same energy it has a γ that is 2000 times larger. The emission of synchrotron radiation depends on γ^4 and it turns out that across the entire window of energy in which we build synchrotrons the γ of electrons is high enough to cause copious emission of radiation in the visible spectrum and at higher frequencies. In contrast, it is only at the highest energies for which proton synchrotrons have been designed that this radiation is at all significant. Unfortunately, all the energy radiated by electrons must be restored to the beam by the r.f. cavities or the particle will slow down and be lost. Electron machines have powerful r.f. systems and much of the voltage per turn they apply is merely to keep the beam from decelerating.

We shall see later that the radiation losses are proportional to the curvature of the path and hence the designer of an electron machine often chooses not to design for the highest magnetic field, which would merely aggravate this curvature. The ring is much larger than the limit imposed by magnetic rigidity in order to keep losses within acceptable bounds. There is little incentive to use superconducting magnets in high-energy electron machines since they only reduce the radius. The extra cost of the longer circumference is more than offset by the reduction of r.f. related costs.

We shall see that synchrotron emission also makes itself felt in determining the energy spread and betatron emittances of the electron beam. The emission of discrete quanta of synchrotron radiation is a random process which continually stirs up the beam, dilating its emittance in all three degrees of freedom.

Fortunately, as we shall discover, higher-momentum particles within the bucket emit much more radiation than their partners at lower energy; a process which, if left to itself, would damp the beam to become monochromatic. The phase-space area of such a beam is no longer determined by Liouville's theorem as it would be in a proton machine, but it results from an equilibrium between these two effects, both of which are consequences of radiation.

8.1.1 Emission of radiation

The physical explanation of the emission of synchrotron radiation deserves a detailed discussion. Further explanation can be found in Appendix 1 and for an even more rigorous and general description the reader is referred to well-known texts (Hübner 1984; Jackson 1962; Walker 1992a,b).

Some readers may be familiar with *bremsstrahlung*, the so-called braking radiation that is emitted as charged particles are slowed down in their interaction with matter. Synchrotron emission is a related phenomenon, but while in the case of *bremsstrahlung* the force is usually applied in opposition to the velocity vector, in a synchrotron the bending force and the acceleration f are both normal to the trajectory. There is nothing esoteric about synchrotron radiation—it is just like the propagating electromagnetic wave set up by a moving charge such as the charges oscillating up and down a dipole transmitting aerial. A classical formula attributed to Larmor, and found in most physics text books, for the power radiated by a moving charge is:

$$P_\gamma = \frac{1}{4\pi\varepsilon_0}\frac{e^2}{c^3}\cdot f^2,$$

where f is the acceleration. For this to apply, the electron and the observer must be moving with a relative velocity which is much less than c. It is not necessary to invoke quantum mechanics to arrive at the above expression; one can apply Maxwell's equations and allow for the finite time for fields generated by the moving charge to reach the observer.

The rigorous derivation for a relativistic particle circulating in a synchrotron can be found in Appendix 1. The result is

$$P_\gamma = \frac{1}{6\pi\varepsilon_0}\frac{e^2 f^2}{c^3}\cdot\gamma^4.$$

Apart from a small numerical factor, the main difference between this and the classical formula is through the quantity γ^4. To understand this, we must examine the Lorentz transformation from the moving frame to the laboratory, where we find that the acceleration normal to the direction of motion contains a factor of γ^2.

8.1.2 Application to a synchrotron

Now we come to apply the expression to a beam circulating in an electron synchrotron to see how it scales with energy and with the radius of curvature. We remember that for motion in a circle f is simply v^2/ρ and since, in all important cases, $v \approx c$,

$$P_\gamma = \frac{1}{6\pi\varepsilon_0}\frac{e^2 c}{\rho^2} \cdot \gamma^4.$$

In order to see how the power scales with the energy of the beam, we may simply substitute $\gamma^2 = E^2/m_0^2c^4$ and use the expression for the classical radius of the electron,

$$r_\mathrm{e} = \frac{e^2}{4\pi\varepsilon_0 m_0 c^2} = 2.8179 \times 10^{-15}\ \mathrm{m};$$

we have

$$P_\gamma = \frac{2}{3}\frac{r_\mathrm{e} c}{(m_0c^2)^3}\frac{E^4}{\rho^2}.$$

One finds that for a large machine like LEP the power can be several MW. This must be pumped into the beam through the r.f. cavities.

The above formula for the power applies to an electron circulating in the synchronous orbit. It shows how the power varies among synchrotrons of different energies and radii assuming that the electron is centred in the vacuum chamber. The situation is different if we fix the mean radius and field of the synchrotron and want to see how particles of different energies radiate. In this case B is constant and ρ is a function of momentum. We can then substitute,

$$\frac{1}{\rho^2} = \frac{B^2e^2}{p^2} = \frac{B^2e^2c^2}{(pc)^2} \approx \frac{B^2e^2c^2}{E^2}$$

and obtain what we call the 'constant field' formula:

$$P_\gamma = \frac{2}{3}\frac{r_\mathrm{e} e^2}{(m_0 c)^3}E^2B^2.$$

8.1.3 The energy lost per turn

The power radiated is important from the r.f. engineering point of view and one must optimize lattice parameters with this in mind. To estimate the total voltage required to keep the beam circulating, we should also know the energy radiated by a particle on each turn. This is obtained by multiplying the power by the revolution time $(2\pi R/\beta c)$:

$$U_0 = \frac{4}{3}\pi\frac{r_\mathrm{e}}{(m_0c^2)^3}\frac{E^4}{\rho}.$$

It can be seen from Table 8.1 that, as higher-energy electron rings become important for particle physics, it becomes difficult to keep U_0 within reasonable bounds by increasing ρ. It is quite natural in a synchrotron to scale U_0 and ρ linearly with energy but in an electron machine there is an additional factor E^3 to be compensated. This becomes impractical at energies above 100 GeV even with a machine like LEP, whose radius of curvature has already been chosen to

Table 8.1 Synchrotron radiation from large synchrotrons

	U_0	$P_\gamma \times N_{\text{beam}}$
LEP at 50 GeV	126 MeV/turn	1.6 MW + 14 MW (ohmic)
LEP at 100 GeV	2.9 GeV/turn	18 MW + 224 MW (ohmic)
500 GeV in 250 km ring	220 GeV/turn	100 MW

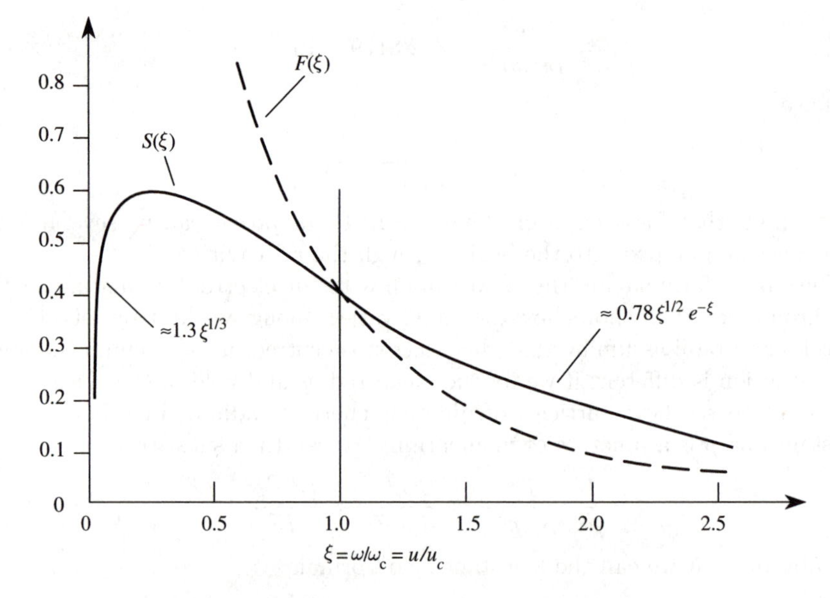

Fig. 8.1 Spectrum of synchrotron emission.

be about 16 times larger than one would normally choose for a proton machine with warm iron magnets.

8.1.4 The spectrum of frequencies

To an observer somewhere on the circumference, a bunch of electrons circulating in a synchrotron will appear as a delta function of charge and hence the spectrum radiated (Fig. 8.1) is broad. This shows a universal curve, normalized by dividing by the energy of a critical quantum:

$$u_{\mathrm{c}} = \hbar\omega_{\mathrm{c}} = \frac{3}{2}\frac{\hbar c\gamma^3}{\rho}.$$

At the same time, the spectral density, $S(\xi) \approx 0.78\xi^{1/2}e^{-\xi}$, is normalized to give a unit integral under the curve. Its shape is independent of both machine parameters and the energy of the beam but, of course, the spectral density and the breadth of the spectrum depend on the energy of a critical quantum.

8.1.5 The rate of emission of quanta

One can get an idea of the number of photons emitted by each electron per second by dividing the power by this critical quantum. More careful averaging gives

$$\mathcal{N} = \frac{15\sqrt{3}}{8}\frac{P_\gamma}{u_c}.$$

8.2 Damping of synchrotron motion

Let us now turn to the virtues rather than the vices of electron machines. We are already familiar with Liouville's theorem which tells us that the normalized emittance in a high-energy synchrotron can never be smaller than the first stage of the linac that feeds it. However, we have learned that this theorem applies only when the system is conservative—when there is no internal damping mechanism. In an electron machine there is such a damping mechanism, provided by the synchrotron radiation emission. The transverse emittance of an electron ring is not subject to Liouville. In fact synchrotron emission 'cools' the beam, not only in the longitudinal phase plane, but also in the transverse plane. At the same time, the quantized nature of the emission of synchrotron light 'warms' the beam and we find the emittances are determined by the equilibrium between heating and cooling in each plane. Within limits we are free to influence the equilibrium value by an appropriate choice of lattice configuration and parameters.

Figure 8.2(a) shows the familiar trajectory of a proton in longitudinal phase space. At the risk of causing confusion by multiple notation, but for purely historical reasons, we use ε for the difference in energy between our test particle and the synchronous particle. In the case of protons, we have ignored the tiny effect of synchrotron emission and we see that the particle, therefore, follows a closed contour.

In Fig. 8.2(b) we see the effect of synchrotron radiation. We are tempted to calculate the energy lost as the particle follows a trajectory in longitudinal phase space with the following expression, which we derived earlier:

$$P_\gamma = \frac{2r_e c}{3(m_0c^2)^3}\frac{E^4}{\rho^2},$$

but this would be wrong for it assumes that $E = ecB\rho$, which may be true for the synchronous particle but not when a particle changes its energy in a fixed guide field. We had better use the 'constant field' formula:

$$P_\gamma = \frac{2}{3}\frac{r_e e^2}{(m_0 c)^3}E^2B^2.$$

However that there is still a quadratic dependence of power on energy, which provides the damping. Of course, we must remember that all the time the energy loss of the reference particle is being made up by the accelerating voltage in

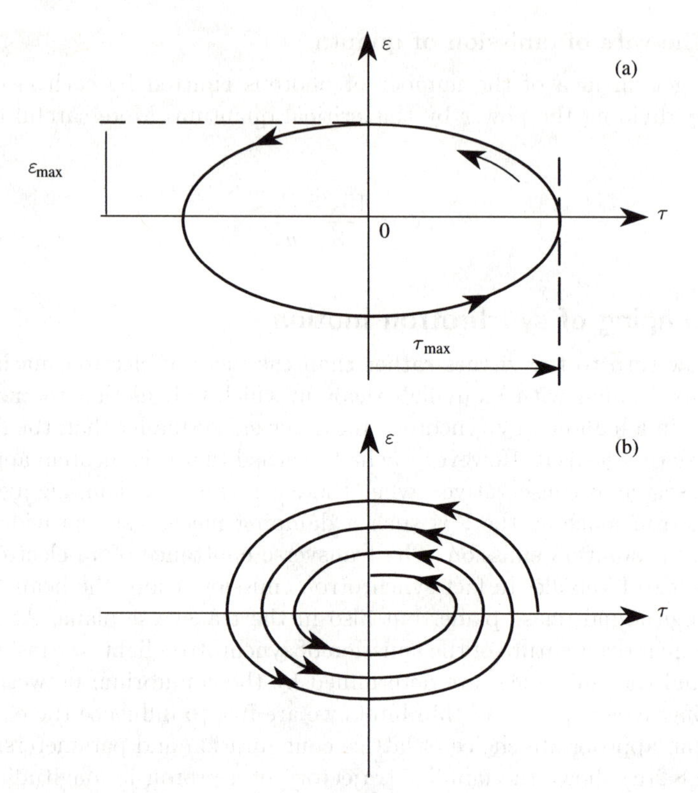

Fig. 8.2 Synchrotron motion in the case of (a) protons and (b) electrons.

the r.f. cavities and that the particle we are tracing also receives the same 'topping up'.

Were it not for the quadratic dependence, both particles would neither lose nor gain energy and the picture would be the same as for protons. However, because of the quadratic dependence, a particle which follows the upper half of the trajectory loses more energy than the reference or synchronous particle and it migrates inwards in the phase diagram towards the origin. In the lower half of the trajectory it also migrates inwards because now it loses less energy relative to the synchronous particle, and the 'topping up' voltage applied indiscriminately to both particles is too generous for the lower-energy electron.

Now let us be more quantitative. The power lost is

$$P_\gamma = \frac{2}{3}\frac{r_e e^2}{(m_0 c)^3} B^2 (E_0 + \varepsilon)^2 \approx \frac{2}{3}\frac{r_e e^2}{(m_0 c)^3} B^2 (E_0^2 + 2E_0 \varepsilon).$$

The first term in the brackets is the energy loss of the synchronous particle which is compensated by the r.f. cavities. The rate of energy loss with respect to the synchronous particle is just the second term, and we can write the differential as

$$\frac{d\varepsilon}{dt} = -\varepsilon\frac{\partial P_\gamma}{\partial \varepsilon} = -\frac{2P_\gamma}{E_0}\varepsilon.$$

8.2.1 Behaviour of an electron with energy defect

As a particle follows the elliptical path of synchrotron motion, it loses more energy on the upper half of the trajectory and gains more on the lower half than does the synchronous particle. It, therefore, follows a spiral path,

$$\varepsilon(t) = \varepsilon_0 e^{\alpha t}\cos\Omega t,$$

and the time constant of the exponential damping is

$$\alpha = \frac{1}{\varepsilon}\left|\frac{d\varepsilon}{dt}\right| = \left\langle\frac{P_\gamma}{E_0}\right\rangle = \frac{1}{\text{Time to lose all its energy}}.$$

We find that the damping time, the reciprocal of α, is simply the time it would take for the particle to lose all its energy (linearly) and may be as short as a few milliseconds. It is easy to imagine how such a fast damping time can dominate the dynamics of an electron machine and make it much less prone to instability. Single-particle instabilities due to high-order non-linear resonances and many collective instabilities with longer growth times, may never get a chance to develop.

The damping time we defined above refers to the linear dimensions of a longitudinal phase-space diagram. We now want to compare this damping rate with the rate of dilation of phase space due to the 'graininess' of emission of quanta of synchrotron radiation. To make this comparison, we must use the damping rate for the *area* of the longitudinal phase ellipse which is twice α.

8.2.2 Quantum emission

There must, of course, be some competing mechanism, otherwise the beam of electrons would shrink rapidly and become perfectly monochromatic (Rees 1989). The mechanism which causes the growth of the energy spread has its origins in the fact that the emission occurs as discrete photons. Each photon takes away some energy, and the point in longitudinal phase space which represents the electron is displaced vertically downwards—a negative increment in ε. At first glance, this looks as if it would have an equal probability of either increasing or reducing the energy defect of the electron with respect to the synchronous particle but let us look closer at the statistical nature of the phenomenon.

To make the problem more tractable, let us choose energy units for the ordinates of our phase-space diagram such that the elliptical trajectory becomes a circle of radius A. We have already established that the average rate of quantum emission is given by

$$\mathcal{N} = \frac{15\sqrt{3}}{8}\frac{P_\gamma}{u_c}.$$

Each quantum reduces the electron energy by an average value of

$$u_{\mathrm{c}} = \hbar\omega_{\mathrm{c}} = \frac{3}{2}\frac{\hbar c\gamma^3}{\rho}.$$

8.2.3 Random walk

If we are allowed to be somewhat cavalier with the averaging process, just to convince ourselves that emission dilates the beam, let us assume that the rate is steady and each quantum has this value. The changes in energy are purely one-dimensional and the problem is a classic random walk of n equal steps of magnitude u_{c} starting from a fixed initial point. The displacement grows as the root of n:

$$\Delta A = \sqrt{n}u_{\mathrm{c}}.$$

The area grows as the square of the amplitude, the rate of change of A^2 being

$$\frac{d\langle A^2\rangle}{dt} = \mathcal{N}\langle u^2\rangle.$$

A more careful integration over distributions of energy and time gives the logarithmic rate of growth of area:

$$\frac{1}{A^2}\frac{d(A^2)}{dt} = \frac{11}{27}\frac{\mathcal{N}\langle u_{\mathrm{c}}^2\rangle}{A^2}.$$

8.2.4 Equilibrium

We can now equate this to the emittance shrinkage rate in order to find the equilibrium value of A. Recall that the growth is in equilibrium with the damping when it is equal to the rate of damping of phase-space area, that is, twice the natural damping rate for amplitude we calculated above:

$$\frac{11}{27}\frac{\mathcal{N}u_{\mathrm{c}}^2}{A^2} = 2\alpha = \frac{4P_\gamma}{E_0}.$$

Solving this to find the equilibrium amplitude, we obtain

$$A^2 = \frac{11}{27}\mathcal{N}u_{\mathrm{c}}^2 \cdot \frac{E_0}{4P_\gamma}.$$

From earlier analysis we know that

$$\mathcal{N}\frac{u_{\mathrm{c}}}{P_\gamma} = \frac{15\sqrt{3}}{8}$$

and

$$u_{\mathrm{c}} = \frac{3}{2}\frac{\hbar c\gamma^3}{\rho},$$

$$A^2 = \frac{11}{27}\cdot\frac{15\sqrt{3}}{8}\cdot\frac{1}{2}\cdot\frac{3}{2}\frac{\hbar\gamma^4}{m_0 c\rho}.$$

The fluctuations are statistical and result in a Gaussian distribution of energy about the mean value. Taking the projected value of A on the energy axis, the σ of the distribution can be calculated and expressed as a relative energy spread:

$$\left(\frac{\sigma_\varepsilon}{E}\right)^2 = \frac{55}{64\sqrt{3}}\frac{\hbar}{m_0 c}\left[\frac{\gamma^2}{\rho}\right]$$
$$= 1.92 \times 10^{-13}\frac{\gamma^2}{\rho}.$$

Having said this, a word of warning to readers who are more familiar with proton synchrotrons: the emission of quanta is continuously stirring up the soup of electrons, and individual electrons from time to time find themselves in the tails of the Gaussian. We shall see later that one may calculate how long it takes for half the beam to have found itself at some time in its history as much as 6σ in energy defect from the centre of the distribution. This is a time of only a few hours and could be the major source of beam attenuation in a storage ring unless we allow a free space about the beam of at least 6σ. In proton machines, the usual allowance is only 2σ excitation of betatron amplitudes.

8.3 Excitation of betatron amplitudes

Quite similar concepts of excitation, damping, and the establishment of an equilibrium between them come into play in the transverse phase plane. Again, the growth is due to quantum graininess. In Fig. 8.3 we see an electron approaching from left to right, making betatron oscillations of amplitude x_β about a closed orbit, which follows the dispersion function $D(s)$ and whose displacement is, therefore, $x_\varepsilon = D(s)\delta p/p \approx D(s)\delta E/E$. When a quantum is emitted, there is a sudden change in energy, which means that the particle begins to oscillate about a different reference orbit. The energy of the photon of synchrotron radiation

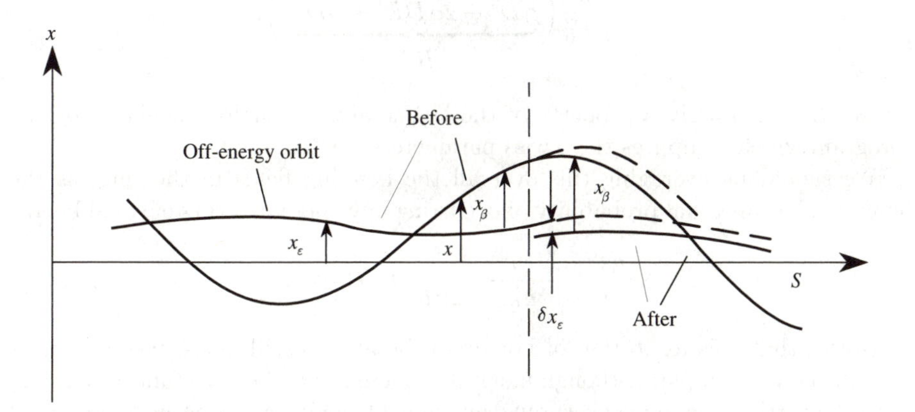

Fig. 8.3 Changes in betatron amplitude at the point of emission.

emitted is u, and from then on the electron's closed orbit must be that of its new energy. This is lower by u, and hence is instantaneously displaced towards the centre of the ring by $\delta x_\varepsilon = D(s)u/E$. Of course, there can be no discontinuity in the displacement or the divergence of the electron because the photon is emitted tangential to its path. The reduction of the displacement due to the energy change must be exactly balanced by an increase in the betatron amplitude to leave the quantity:

$$x_\beta + D(s)\frac{\delta E}{E}$$

unchanged. Hence, we have the important physical link between the transverse plane and the energy loss. The increment to its betatron amplitude is just the fractional energy change multiplied by the local value of the dispersion:

$$\delta x = -D\frac{u}{E_0}.$$

8.3.1 The effect on the emittance

Without enquiring too closely into the rigour of the expression, we can imagine that the increase in the horizontal emittance or, for a single particle, of the Courant and Snyder invariant, is

$$\delta\varepsilon = \delta(x^2) = \frac{u^2 D^2}{E_0^2}.$$

Here we are using the standard relation between the square of the displacement, betatron amplitude, and emittance (the other use of the symbol ε) together with our notion of a random walk. With a more care we may obtain

$$\begin{aligned}\delta\varepsilon &= \delta(\gamma x^2 + 2\alpha x x' + \beta x'^2) \\ &= \frac{u^2}{E_0^2}\Bigg(\underbrace{\gamma D^2 + 2\alpha D D' + \beta D'^2}_{H(s)}\Bigg),\end{aligned}$$

where $H(s)$ is purely a property of the lattice and is usually calculated by the program which computes the Twiss parameters.

We should be averaging this over all the bending fields in the ring, as the emission has an equal probability of occurring anywhere in a ring of equal bends:

$$\frac{d\varepsilon}{dt} = \frac{\mathcal{N}\langle u^2\rangle}{E_0^2}\cdot\frac{1}{2\pi R}\int H(s)\,ds.$$

Notice that this excitation of horizontal betatron oscillations has no equivalent in the vertical plane, though inevitably there is a certain amount of coupling into the vertical plane via skew quadrupole fields which contributes to a vertical beam size.

8.4 Damping of betatron oscillations

Now we must examine the damping mechanism which, with the excitation, results in an equilibrium beam width. The photon emitted by the electron is co-linear with the path of the electron within a very small angle which is of the order $1/\gamma$ and quantum emission does not change the local displacement and divergence. Figure 8.4 shows the small resultant change $\delta\mathbf{p}$ in the momentum vector. At first sight this does not seem to excite or damp the betatron motion, at least in the vertical plane which we portray.

The damping mechanism arises, not because of what happens at the site of emission, but because of the inability of the r.f. cavity to restore $\delta\mathbf{p}$ exactly.

As soon as the electron loses energy, the principle of phase stability comes into play so that the electron arrives at the r.f. cavity at a phase where it is given a little extra energy to correct for the loss. But the r.f. cavity can only accelerate the electron in the longitudinal direction, increasing $p_{\parallel}$ without affecting $p_{\perp}$ (Fig. 8.5).

If we reflect carefully, we can see that because the emission always reduces the electron's energy, the effect of the cavity can only be to decrease the divergence,

$$z' = \frac{p_{\perp}}{p_{\parallel}} \rightarrow \frac{p_{\perp}}{p_{\parallel} + \delta p} \approx z'\left(1 - \frac{\delta p}{p}\right),$$

leading to a steady damping of the betatron motion.

At first glance, the fractional change in divergence just seems to be the fractional change in momentum and the damping time, the same as that for energy.

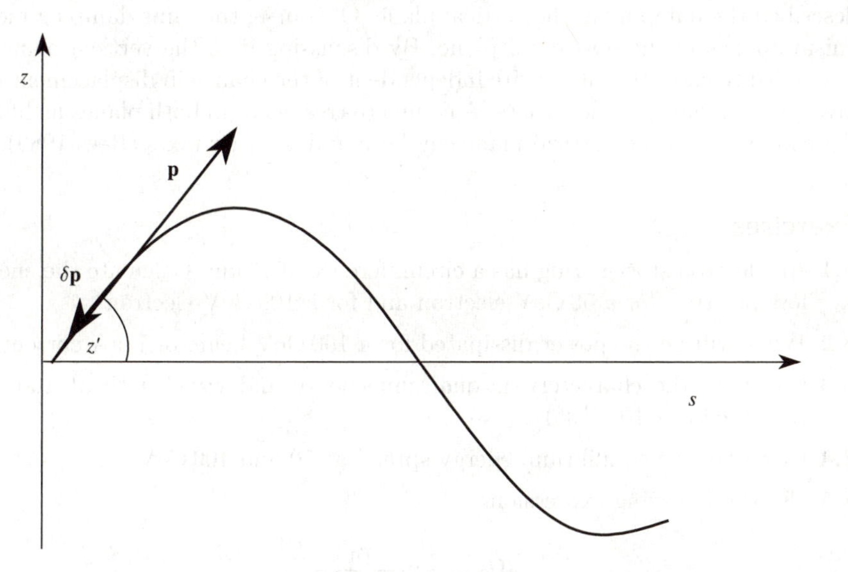

Fig. 8.4 Change in momentum at the point of emission.

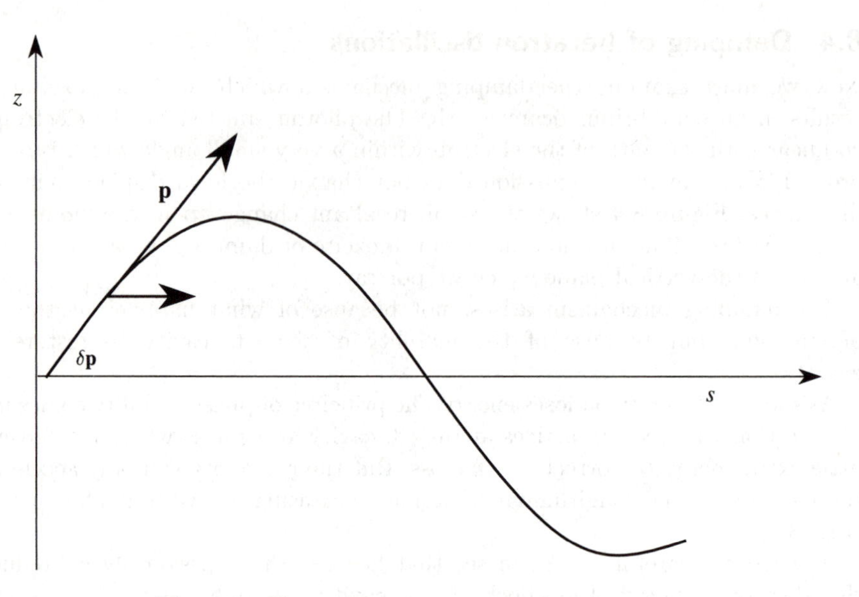

Fig. 8.5 Change in momentum at the r.f. cavity.

However, when we look at the average effect projected on the displacement axis, the rate turns out to be half of that for energy.

8.4.1 The vertical plane

The observant reader would not have failed to notice that we have chosen to explain the excitation of betatron oscillations in the horizontal plane but described the damping in the vertical plane. Of course, the same damping mechanism applies in the horizontal plane. By discussing it in the vertical plane we have tried to show that it is quite independent of the change in displacement and divergence due to the energy loss. A complete treatment in both planes including the coupling into the vertical plane can be found in other texts (Rees 1989).

Exercises

8.1 An electron storage ring has a circumference of 27 km. Calculate the energy loss per turn for a 50 GeV electron and for a 100 GeV electron.

8.2 What will be the power dissipated for a 100 GeV beam of 1 mA current?

8.3 Calculate the characteristic quantum energy and wavelength at 100 GeV ($h = 6.6262 \times 10^{-34}$ Js)

8.4 Calculate the equilibrium energy spread at 50 and 100 GeV.

8.5 Use the following expression,

$$U_0 = \frac{4}{3}\pi \frac{r_0}{(m_0c^2)^3}\frac{E^4}{\rho},$$

to calculate the energy loss per turn of a 4 TeV muon in a ring of 4 km radius (muon mass is 105 MeV).

8.6 What would be the characteristic frequency of the synchrotron light emitted?

8.7 Calculate the damping time.

9 Space charge and instabilities

9.1 Transverse space charge

Figure 9.1 shows a beam of cylindrical cross section. A proton located at (r, ϕ) experiences only the electrostatic fields from its neighbours:

$$E_r^* = \frac{\rho^*}{2\varepsilon_0} r^*,$$

and a radial defocusing force

$$f^* = eE^*,$$

where the asterisk indicates the coordinate system moving with the proton.

Transforming this relativistically into the laboratory system, the protons behave as line currents, producing magnetic fields and a mutually repelling radial

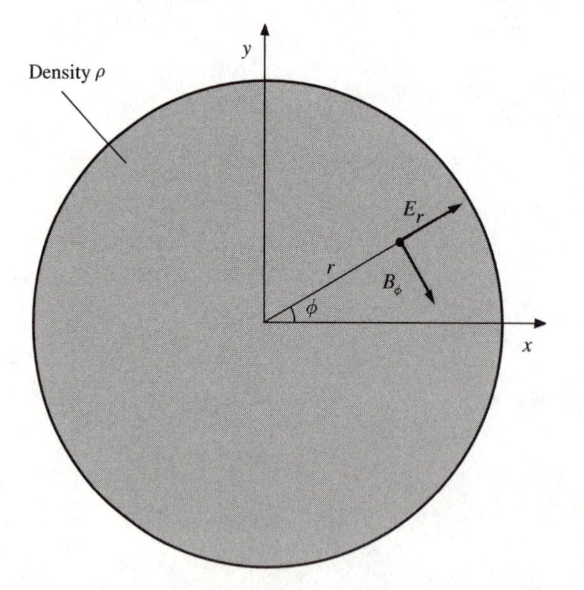

Fig. 9.1 Space-charge fields in a cylindrical beam of uniform density.

defocusing force:

$$E_r = \frac{\rho}{2\varepsilon_0} r, \qquad B_\phi = \frac{\rho}{2\varepsilon_0} \frac{v}{c^2} r,$$

$$\delta f_r = e(\bar{E} + \bar{v} \times \bar{B}) = \frac{e\rho}{2\varepsilon_0}(1 - \beta^2)r = \frac{e\rho r}{2\varepsilon_0 \gamma^2}.$$

We can equate this to the rate of change of transverse momentum, velocity and, finally, the divergence:

$$\frac{e\rho r}{2\varepsilon_0 \gamma^2} = \delta f_r = \frac{d}{dt}(p_{\mathrm{T}}) = \frac{d}{dt}\left(\frac{m v_{\mathrm{T}}}{\sqrt{1-\beta^2}}\right)$$

$$= m\gamma \frac{d^2 r}{dt^2} = m\gamma(\beta c)^2 \frac{d^2 r}{ds^2}.$$

Using the classical proton radius

$$r_0 = \frac{e^2}{4\pi\varepsilon_0 m_0 c^2} = 1.5 \times 10^{-18}\ \mathrm{m},$$

to mop up a lot of constants, we arrive at

$$\frac{d^2 r}{ds^2} = \left\{\frac{r_0 N}{\beta^2 \gamma^3 RS}\right\} r,$$

where R is the radius of machine, S the beam cross-section, and N the number of circulating protons.

Remembering that r can be either x or z in our Cartesian system for describing betatron motion, we see that the equation is none other than Hill's equation with k a defocusing term acting all round the circumference,

$$k = \frac{-r_0 N}{\beta^2 \gamma^3 RS}.$$

As a consequence, in both the horizontal and the vertical planes, the Q is shifted downwards by an amount which can be calculated from the previous formulae for the effect of gradient errors in Chapter 6 assuming the mean betatron function is R/Q:

$$\delta Q \approx \frac{-r_0 RN}{2Q\beta^2\gamma^3 S}.$$

At first sight, this seems not to be a nuisance. Uncorrected, a δQ of a few times 10^{-2} can be accommodated between the stopbands and, if it gets larger, retuning the lattice quadrupole strength will restore the working point. But there is a limit to how far one can apply such compensation. In practice, for accelerators this is usually taken as $\delta Q < 0.25$. The difficulty is that protons near the edge of the r.f. bunch find themselves, twice every synchrotron oscillation, in a region of rarefied density at the head and tail of the bunch. The dQ is not the same for all protons or even for the same proton at different points in the synchrotron motion.

The intensity N, which gives a δQ of -0.25, is usually quoted as the *space-charge limit.* Since δQ is strongly dependent on β and γ and is most serious at injection, the intensity limit it imposes can be considerably improved if a booster synchrotron is used to raise the injection energy. It used to be about $N = 10^{12}$ for the CERN-PS until a booster was inserted to raise the energy of injection from 50 to 800 MeV.

Readers are warned not to rely on the above formula for δQ in numerical calculations. The full and accurate formula can be found in Laslett and Resegotti (1967) and Hofmann (1992).

The more accurate formula takes into account that

(i) the beam is bunched and, therefore, concentrated by a factor of $B = (\phi_2 - \phi_1)/2\pi$;
(ii) the image forces in the walls of the vacuum chamber play an important role in determining the absolute dQ. In fact, the vacuum chamber size is more important than S at high energy.

9.2 General features of collective instabilities

Once the trivial hardware faults are eliminated and an accelerator is brought into operation with a circulating beam, we may expect the intensity to improve as corrections are applied to compensate magnetic imperfections, the optics is tuned to design values and non-linear behaviour compensated. Yet often, just as the milestone of achieving the design intensity is in sight, the beam suddenly disappears as if there were some critical intensity which cannot be exceeded. This can be a symptom of collective instability in which the charge of the beam itself produces an oscillating field strong enough to perturb the density distribution or displacement of another part of the circulating beam. If the phase relation between the field and the perturbation of density is in the right direction, it will reinforce the oscillating field and the beam will behave like an amplifier with positive feedback.

We will not try to describe in detail the many different kinds of instabilities. They may be caused by perturbing forces acting in either the longitudinal or transverse phase planes which are often produced by resonating cavities or other objects in the ring. In some cases, they can link the head to the tail of a single bunch or couple oscillations in the centre of charge of one bunch to the motion of another bunch. In this chapter, which is only an introduction to a vast subject, we take only the simple example of a longitudinal instability excited by a resonant cavity. However, before we study this, let us look at an even simpler instability which does not require any resonant cavity but illustrates the important point of how the conditions for instability change as we cross transition.

9.3 Negative-mass instability

Although this instability is not one of the most troublesome encountered in modern synchrotrons, it is an excellent starting point in the understanding of collective effects and how they grow.

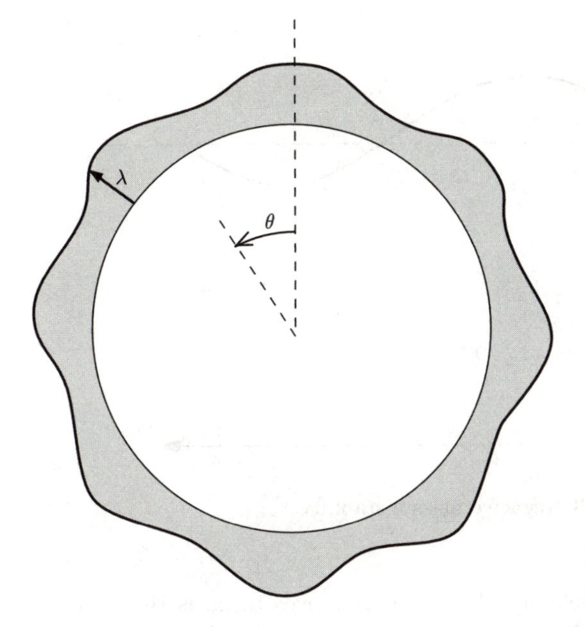

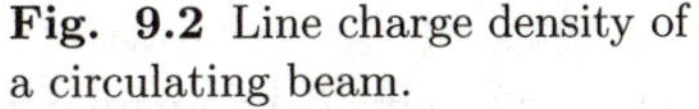

Fig. 9.2 Line charge density of a circulating beam.

Figure 9.2 shows how the local line density of charge might vary around a synchrotron once the instability begins to develop. One of the tricks we use in understanding an instability is to postulate such a simple form for the perturbation, analyse its effect on the beam distribution, and then test whether such a pattern is self-sustaining. If the forces due to the field setup reinforce its shape, it will grow exponentially from a random component of noise in the azimuthal variation of λ.

The azimuthal modulation is replotted in Cartesian form in Fig. 9.3. We first examine the particle at A. This finds itself with a larger charge density behind it, pushing it forward, while particle B will be decelerated by the mountain of charge in front of it. The field due to this space-charge force may be written as

$$E = -\left[\frac{e}{4\pi\varepsilon_0\gamma^2}\right]\frac{\partial\lambda}{\partial s}.$$

The quantity in square brackets is a constant which includes the $1/\gamma^2$ factor to be expected when we transform a purely electrostatic space-charge effect to the laboratory frame in which we express the field. However, let us not become mesmerized by relativity. The important physics is in the term $\partial\lambda/\partial s$. This indicates that if the effect of the field on the particle distribution is to make $\partial\lambda/\partial s$ larger, this will reinforce E and we may expect unstable growth.

At first sight, this appears not to be the case as we find that the particle A is accelerated away from the hump and B decelerated backwards, thus flattening out the perturbation. This is the picture below transition but we must not forget that, above transition, accelerating a charge at A will not significantly increase its speed but, more importantly, cause it to take a circular orbit of larger radius.

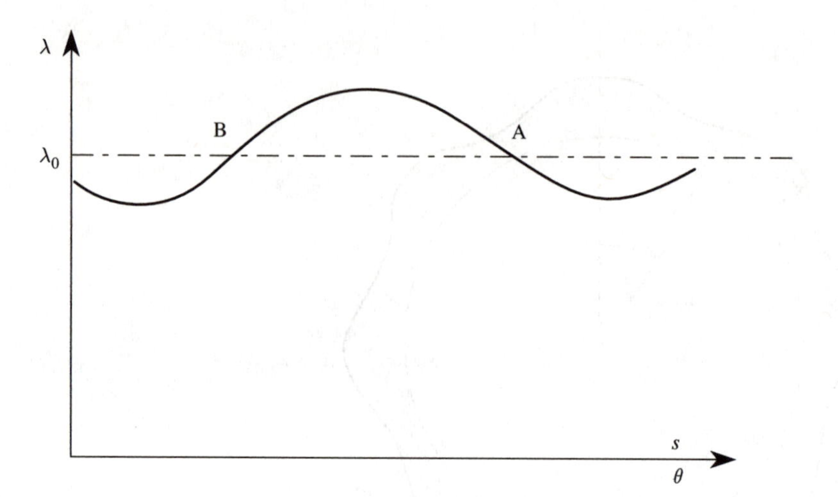

Fig. 9.3 Negative mass instability.

This will reduce A's angular velocity, $\dot{\theta}$, so that A moves towards the summit. Similarly, as B is 'decelerated' its angular velocity increases to join the summit. Hence, as we pass through transition we enter a new regime where this kind of 'negative mass' instability can grow from noise. The reader may wonder how any synchrotron can overcome this barrier and certainly this was a concern of the builders of both AGS and CERN-PS, the first strong-focusing synchrotrons, as they pushed up their intensities. Later in this chapter we will discuss a mechanism of stabilizing or damping the growth of such an instability.

9.4 Longitudinal instability

9.4.1 Driving terms

In this study of instabilities we must first introduce a number of individual concepts before we are able to piece the theory together. The first of these is the concept of self-reinforcing growth discussed above. The second is Fourier analysis of the periodic signal of the bunch as it circulates around the machine (Fig. 9.4). In the frequency domain such a delta function produces a fundamental at the machine's revolution frequency and all its higher harmonics which are present with equal amplitudes,

$$I = \sum I_n e^{-in\omega_0 t}.$$

Figure 9.5 shows this comb of frequencies and, superimposed, the response of an r.f. cavity somewhere in the ring. It only needs one of the frequencies to fall in the response curve of the cavity for it to become excited. The next step in understanding instabilities is to calculate the fields such a bunch can set up in

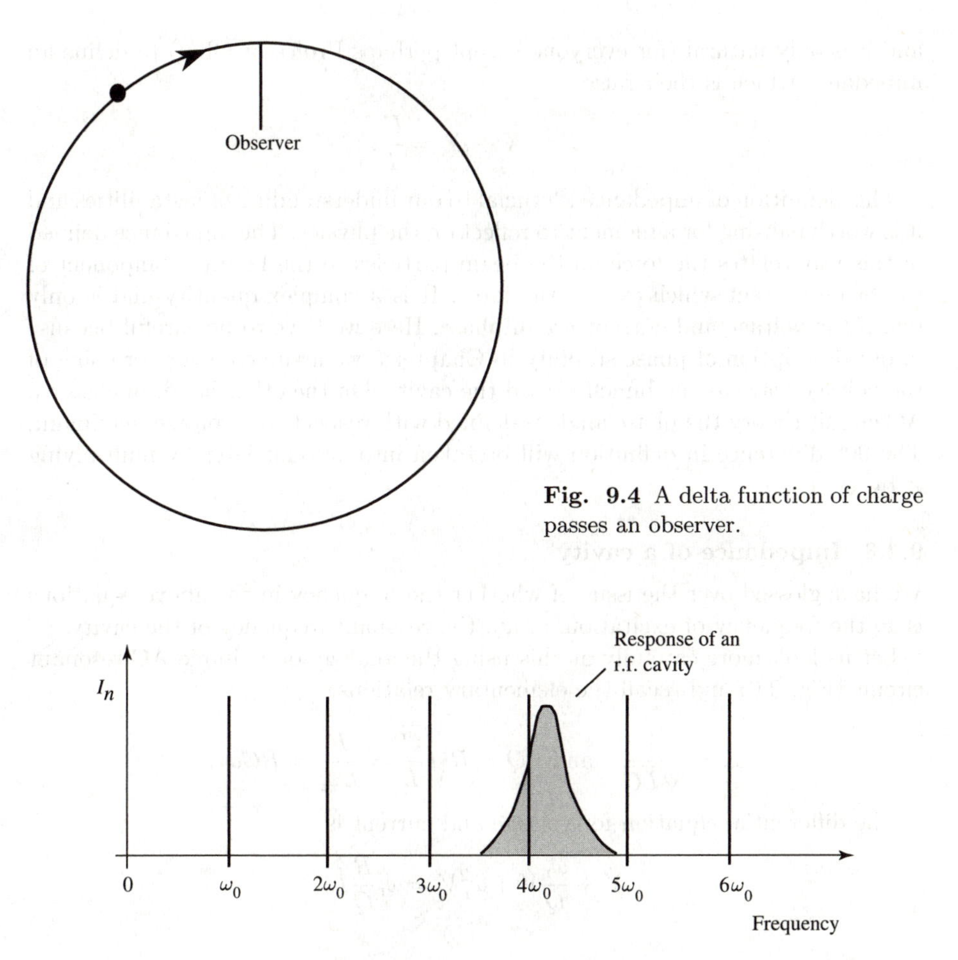

Fig. 9.4 A delta function of charge passes an observer.

Fig. 9.5 Spectrum from a bunch.

a passive cavity. It is these fields, acting back on the beam distribution, which cause the instability to grow.

9.4.2 Exciting a cavity-like object

The cavity we consider need not be one of the r.f. cavities deliberately installed to accelerate particles but any box or local enlargement in the beam tube which can resonate. Figure 1.13 shows such a cavity-like object. Quite generally, we can write the voltage, U, experienced by a particle in the cavity (approximately, the voltage from one end to the other) in the same form as the current which excites it:

$$I = \hat{I}e^{-i\omega t} \quad \text{and} \quad U = \hat{U}e^{-i\omega t},$$

and it is only natural (for everyone except perhaps Professor Ohm) to define an impedance which is their ratio:

$$Z = X + iY = \frac{U}{I}.$$

This definition of impedance is crucial to our understanding of instabilities and it is worth pausing for a moment to reflect on the physics. The impedance defined in this way relates the force on the beam particles to the Fourier component of the beam current which excites the force. It is a complex quantity and is only real if the voltage and current are in phase. Here we have to be careful because in our description of phase stability in Chapter 5 we assumed a zero-crossing of the voltage wave as the bunch passed the cavity. On the other hand, in classical AC circuit theory the phase angle is defined with respect to a voltage maximum. The 90° difference in definition will be taken into account later by multiplying Z by $\sqrt{-1} = i$.

9.4.3 Impedance of a cavity

We have glossed over the issue of whether the frequency in the above equations is ω, the frequency of excitation, or ω_r, the resonant frequency of the cavity.

Let us look more carefully at this using the analogy of a simple AC resonant circuit (Fig. 9.6) and recall the elementary relations:

$$\omega_r = \frac{1}{\sqrt{LC}} \quad \text{and} \quad Q = R\sqrt{\frac{C}{L}} = \frac{R}{L\omega_r} = RC\omega_r.$$

The differential equation for voltage and current is

$$\ddot{V} + \frac{\omega_r}{Q}\dot{V} + \omega_r^2 V = \omega_r \frac{R}{Q}\dot{I},$$

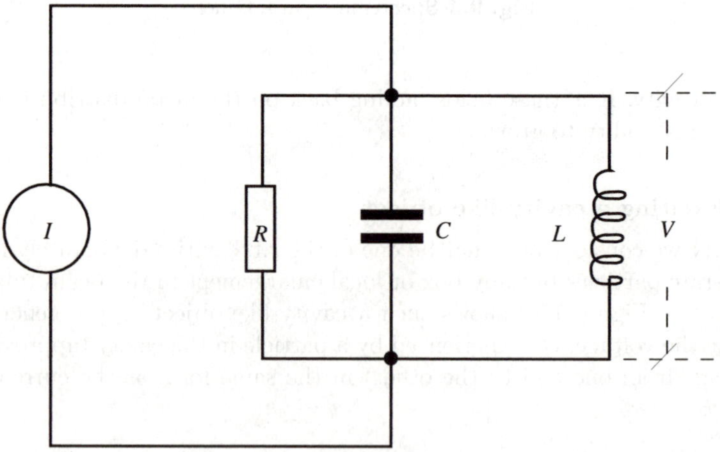

Fig. 9.6 AC resonant circuit.

with the solution

$$V = V_0 e^{-at} \cos\left[\omega_{\mathrm{r}}\sqrt{1-\frac{1}{4Q^2}}t+\phi\right].$$

The damping coefficient

$$\alpha = \frac{\omega_{\mathrm{r}}}{2Q}$$

is the damping rate or, alternatively, the reciprocal of the decay time.

Text books on AC circuit theory show that if $I = \hat{I}e^{-i\omega t}$ then the impedance seen by the generator is

$$Z(\omega) = R\left[\frac{1+iQ(\omega^2-\omega_{\mathrm{r}}^2)/\omega\omega_{\mathrm{r}}}{1+Q^2((\omega^2-\omega_{\mathrm{r}}^2)/\omega\omega_{\mathrm{r}})^2}\right].$$

We write this as $Z = X + iY$ and we can see that when ω is less than ω_{r} the reactive component, Y, is capacitive or negative while, after crossing zero at $\omega = \omega_{\mathrm{r}}$, it becomes positive and inductive as the driving frequency rises above the natural resonance of the oscillator.

9.4.4 Synthesis of the effect of a cavity

Now all the preparatory work is complete and we will soon be able to draw the analogy between the effect of a cavity and the negative-mass instability. We return to Fig. 9.2 and, instead of the line density λ, we consider the beam current

$$I = I_0 + I_1 e^{i(n\theta-\Omega t)}.$$

Here n is the number of humps in the pattern around the ring, while Ω is simply the angular frequency which an antenna in the wall of the vacuum chamber would pick up. We may refer to n as the mode number of the instability. Note that such a pattern is not frozen but will precess. We guess from our experience with negative-mass instability that it is the gradient dI/ds which drives an instability and, ignoring I_0, we write

$$I = I_1 e^{i(n\theta-\Omega t)}.$$

Then for a ring of mean radius R,

$$\frac{dI}{ds} = \frac{1}{R}\cdot\frac{dI}{d\theta} = I_1\frac{in}{R}e^{i(n\theta-\Omega t)} \quad \text{and} \quad \frac{dI}{ds} = \frac{in}{R}\cdot I.$$

We rearrange this as

$$I_1 e^{i(n\theta-\Omega t)} = \frac{R}{in}\frac{dI}{ds}.$$

This seems a rather obscure relation but it enables us to substitute the gradient of the current distribution for the current, I_1, in the next expression and draw an analogy with the negative-mass instability.

According to our definition of impedance, a cavity-like object (at $\theta = 0$) which presents an impedance $X + iY$ at this frequency will produce an accelerating voltage:

$$(X+iY)I_1 e^{-\Omega t} = R\Big(\frac{X+iY}{in}\Big)\frac{dI}{ds}$$
$$= -\frac{iR}{n}(X+iY)\frac{dI}{ds} = \frac{R}{n}(Y-iX)\frac{dI}{ds}.$$

This can be compared with the accelerating field for the negative-mass instability:

$$-\Big(\frac{e}{4\pi\varepsilon_0\gamma^2}\Big)\frac{d\lambda}{ds},$$

which, as we argued before, is unstable above transition and stable below it—a result of the negative sign. The field due to the cavity has no minus sign before it but when Y is negative, that is, capacitive, the right-hand side will become negative and the cavity will have the same effect as negative mass on promoting an instability—it will be unstable above the transition and stable below.

On the other hand, if we think through the logic of the negative-mass instability again we will find that below transition an inductive impedance, for which Y is positive, will decelerate the particle A, on the falling slope of the current wave while accelerating B. This will enhance the 'hump' and cause instability in a region where the negative-mass instability is stable. Just to complete the picture an inductive impedance though unstable below will be stable above transition.

We can approximate the total effect of all the rather low-Q resonators which comprise the large number of small changes in vacuum chamber cross-section by a single cavity with $Q \approx 1$ and with ω_{r} close to the cutoff frequency of the beam pipe (typically 1 MHz). Such an impedance is inductive in its response to beam structure at lower frequencies than 1 MHz and hence is quite liable to provoke an instability below transition.

9.4.5 A short cut to calculating an instability

The accelerating voltage due to the impedance of a cavity is no different in its effect from the voltage imposed on the cavities installed to accelerate the beam. This allows us to take a short cut around complex theory and arrive at a good physical appreciation of instabilities by using the expressions we have already developed for synchrotron motion. We recall the effect of a cavity voltage on the longitudinal motion of a particle. The full equation describing the motion is

$$\frac{d}{dt}\left[\frac{E_0\beta^2\gamma\dot{\phi}}{2\pi\eta h f^2}\right] + eV_0(\sin\phi - \sin\phi_{\mathrm{s}}) = 0.$$

However, for the purpose of demonstrating the growth of an instability, let us first assume that the particles have initially a small phase excursion about

$\phi_S = 0$, so that we may write

$$\left[\frac{E_0\beta^2\gamma}{2\pi\eta h f^2}\right]\ddot{\phi} + eV_0\phi = 0$$

or

$$\ddot{\phi} + \Omega_s^2\phi = 0,$$

where

$$\Omega_s^2 = \left[\frac{\eta h e V_0}{2\pi E_0 \beta^2 \gamma}\right]\omega_0^2,$$

is the synchrotron frequency and ω_0 is the angular-revolution frequency.

Here we take an imaginative leap and courageously replace $V_0 h$, the voltage of an r.f. cavity multiplied by its harmonic number, by

$$inZI_0.$$

We have already defined Z in such a way that ZI_0 is the voltage reacting on the beam, due to current I_0. In addition, we have included n as the effective harmonic number (ω/ω_0) and i to reflect the 90° difference in the definition of zero phase between AC circuit theory and cavities for acceleration. The next step is to put this voltage as a driving term or force on the right-hand side of the differential equation for ϕ above.

The reader may have reservations about this intuitive step yet it bypasses many pages of analysis, gives a clear physical insight into the effect of a passive cavity on the beam and leads immediately to the correct formula for the 'frequency shift' from an impedance Z. More rigorous derivations can be found in Chao (1993), Hofmann (1996), Laclare (1980, 1982, 1992), Sacherer (1972) and Zotter (1976).

9.4.6 Effect of frequency shift

By now, the reader may be used to the concept of impedance we are using but now we come across the term 'frequency shift' which can often also be a stumbling block in our physical understanding. It may help to remember that a force (in this case a voltage), driving an oscillation close to its natural frequency and written as a term on the right-hand side of the differential equation for phase oscillations:

$$\ddot{\phi} + \Omega_0^2\phi = F(t)$$

can be incorporated in the left-hand side as a change in frequency, strictly a change in frequency squared:

$$\ddot{\phi} + (\Omega_0 + \Delta\Omega)^2\phi = 0.$$

We can drop Ω_0 if the r.f. is off.

By incorporating the driving term in this way, we can define a change in the synchrotron frequency (squared) due to the beam induced voltage:

$$(\Delta\Omega)^2 = i\left[\frac{\eta\Omega_0^2 I_0}{2\pi\beta^2 E}\right]\frac{Z}{n} = i\xi\frac{Z}{n}.$$

Here ξ is just to replace the rather complicated square brackets. First let us look at the capacitive (negative Y) case which we found, like the negative-mass instability, was stable below transition (when η is positive). Indeed, $(\Delta\Omega)^2$ being positive there is merely a real shift in frequency. Above transition, $(\Delta\Omega)^2$ will be negative and the frequency shift imaginary, and exponential growth can occur. The argument is reversed as expected for a positive (inductive) Y.

However, even in this 'stable' case, as soon as Z has a resistive component, imaginary terms appear in $\sqrt{iZ}$ and in $\Delta\Omega$ leading to unstable and exponential growth.

Another way to understand this is to write down the solution of the differential equation as

$$\phi = \phi_0 e^{-i\Omega t} = \phi_0 e^{-i(\alpha+i\beta)t} = \phi_0 e^{\beta t} e^{-i\alpha t}.$$

The imaginary frequency is β and this, multiplied by i and then $-i$, produces positive exponential growth. But we must not fall into the trap of expecting a simple relation between the imaginary frequency shift and the resistive component X. The complex frequency shift $\alpha + i\beta$ is not just proportional to $X + iY$ but above transition is proportional to $\sqrt{-i(X+iY)}$.

To complete the understanding, it is best to go through the mathematical analysis, solving for β in terms of only X and Y. There will be combinations of X and Y which lead to a particular value of the growth rate β. This will produce a contour map plotted in X and Y coordinates with β as the height of the mountain. Following any contour line, we find that β, the imaginary part of $\Delta\Omega$, is constant.

The solution above transition where $\eta < 0$ is as follows:

$$\Delta(\Omega)^2 = (\alpha + i\beta)^2,$$

$$\Delta(\Omega)^2 = (\alpha^2 - \beta^2) + 2i\alpha\beta,$$

$$\Delta(\Omega)^2 = -i\xi Z = \xi Y - i\xi X.$$

We now equate the real and imaginary parts:

$$(\alpha^2 - \beta^2) = \xi Y,$$

$$\xi X = -2\alpha\beta,$$

which may be rearranged as

$$\alpha = -\frac{\xi X}{2\beta}.$$

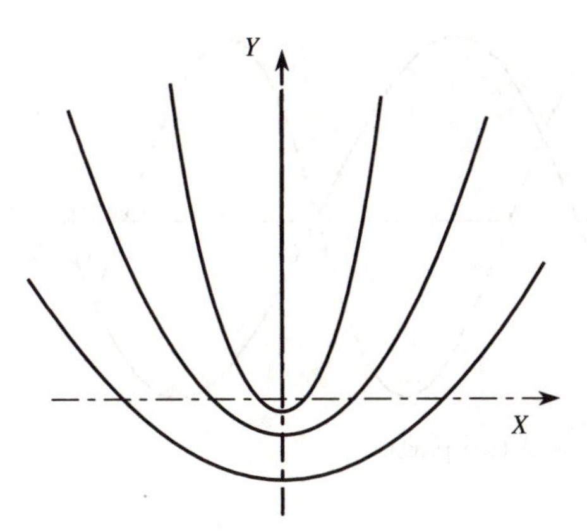

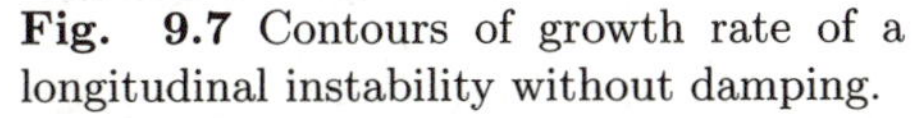

Fig. 9.7 Contours of growth rate of a longitudinal instability without damping.

The next step is to eliminate α by substituting the right-hand side of this expression in the first expression, which gives

$$\xi Y = \frac{\xi^2 X^2}{(2\beta)^2} - \beta^2$$

and then

$$X = 2\beta\sqrt{\frac{Y}{\xi} + \frac{\beta^2}{\xi^2}}.$$

This equation describes contours in the (X, Y) diagram, which are a series of parabolas with growth rate β as their parameter. They are symmetric about the Y-axis and with a minimum at $Y = -\beta^2/\xi$ (Fig. 9.7).

The breadth of the parabolas increases with β, representing stronger and stronger growth as the beam impedance Z increases. The $\beta = 0$ contour, where there is no growth, is simply an infinitely narrow parabola enclosing the Y-axis from the origin to the positive infinity. Thus, above transition, only a purely inductive impedance with absolutely no resistive component will be stable. Of course, below transition, the figure is the same but the positive Y-axis corresponds to a purely capacitive impedance. Capacitive impedance can then be unstable in the presence of a resistive term. A small amount of resistance will make any impedance unstable either above or below the transition. Fortunately, the picture becomes less frightening when we include Landau damping.

9.5 Landau damping

The above treatment and the diagram make an important assumption that all the particles in the beam oscillate with the same frequency Ω_0. However, if this condition is relaxed to allow for a spread in natural oscillation frequency due to, say, momentum spread in the beam, instability is suppressed for small values of β. Any collective motion becomes confused because of the frequency spread

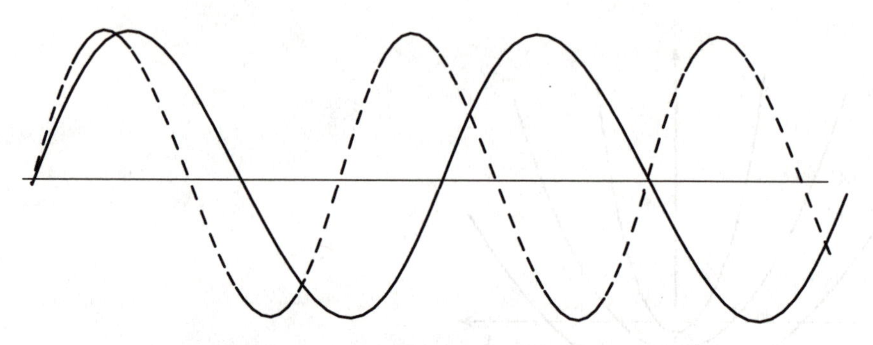

Fig. 9.8 Decoherence of two particles.

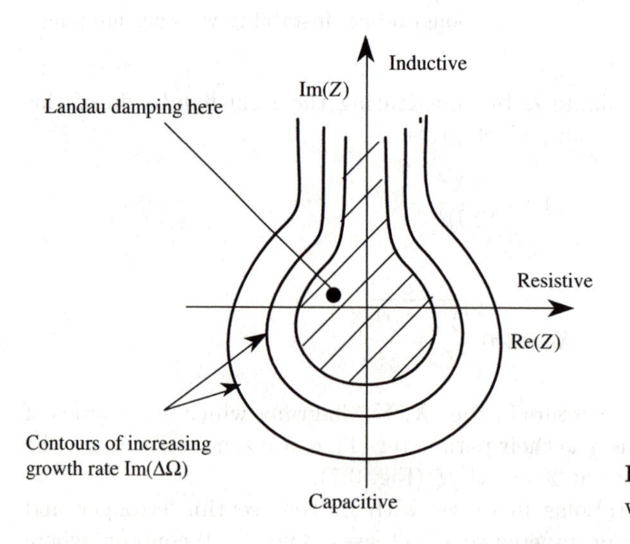

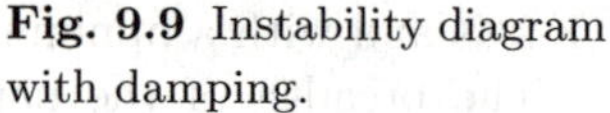

Fig. 9.9 Instability diagram with damping.

on a time scale which is shorter than the time it takes to develop instability (Hereward 1976). Remember that we rely upon the motion of the centre of charge of all the particles to excite the cavity. Figure 9.8 illustrates how two particles of different revolution frequency get out of step.

The Landau damping effect considerably modifies the stability diagram. In general, the contours close to the positive Y-axis are stable, or too weak to grow in competition with the decoherence. We can think of the shaded area in Fig. 9.9 as a stable region. Thus, provided the impedance is inductive a modest resistive component is tolerable, but beyond a certain X value a threshold in stability is passed and growth occurs at a rate characterized by β.

Very often, when an instability affects the beam, one can watch the amplitude of the longitudinal motion grow by connecting an r.f. phase pickup to an oscilloscope. The rise time of the signal gives us $1/\beta$ directly and we can then make an estimate of the modulus of Z and how much frequency spread would be needed to damp the growth.

9.5.1 Keil–Schnell criterion

We can get a rough idea of the threshold for instabilities by equating the frequency spread in the beam,

$$\Delta\Omega = |\eta|\Omega\frac{\Delta p}{p},$$

to the frequency shift $\Delta\Omega$ due to the forces driving the instability,

$$\Delta\Omega \approx \left(\frac{\eta\Omega_0^2 I_0 Z}{2\pi\beta^2 E n}\right)^{1/2}.$$

This gives us a rule of thumb for longitudinal instability; in Europe, it is termed the Keil–Schnell criterion (1969):

$$\left|\frac{Z}{n}\right| \leq \frac{F m_0 c^2 \beta^2 \gamma \eta}{I_0}\left(\frac{\Delta p}{p}\right)^2,$$

where F is a form factor close to unity.

Finally, as we warned the reader, this short cut to understanding longitudinal instability does not pretend to treat the multitude of transverse, head–tail, and multibunch instabilities which can affect the beam. For these the reader should consult Chao (1993) and the other studies referred to in this chapter.

Exercises

9.1 A proton beam of 10^{13} protons is injected into a synchrotron of radius 1100 m at 10 GeV. The normalized emittance in each plane is 10π mm mrad. The Q of the machine is 27.4. The average betatron amplitude function is 60 m. Calculate the ΔQ due to space charge ($r_{\mathrm{p}} = 1.534 \times 10^{-18}$).

9.2 The transition energy for the above accelerator is 24. Calculate the frequency shift for an inductive impedance $Z/n \sim 50\,\Omega$ at 500 MHz.

9.3 What is the growth rate of the longitudinal instability from this frequency shift without Landau damping?

9.4 Plot a parabola in x, y space corresponding to this growth rate.

9.5 Plot parabolas for half and double this growth rate.

9.6 What happens to this diagram above transition?

9.7 Use the Keil–Schnell criterion to calculate a momentum spread which will stabilize the beam for $Z/n = 50\,\Omega$ (assume $F = 1$).

10

Radiofrequency cavities

10.1 Waves and modes in guides and cavities

At the beginning of the 21st century we are at a turning point in the history of accelerators. Circular machines which have dominated the high-energy collider landscape are giving way to linear colliders. We, therefore, devote this chapter to a detailed study of the accelerating cavities themselves.

10.1.1 Necessary conditions for acceleration

In Chapter 1 we saw how in the early 1930s two basic configurations of accelerator were invented: the linear accelerator, or linac, conceived by Ising and Wideröe, and the cyclotron, invented by Lawrence. Both of these use electromagnetic fields oscillating in resonant cavities to apply the accelerating force. In the linac configuration, the particle follows a straight path through a series of cavities. In a cyclotron, and its later development the synchrotron, the beam follows a circular path in a magnetic field and the particles return to the same accelerating cavity each time round. Both these basic configurations make use of oscillating fields and it is instructive to reflect on why this is so. First, we consider Maxwell's equation

$$\nabla \times \mathbf{E} = -\frac{\partial \mathbf{B}}{\partial t}.$$

In its integral form, this becomes

$$\oint \mathbf{E} \cdot d\mathbf{s} = -\frac{d}{dt}\int_S \mathbf{B} \cdot \mathbf{n}\, da.$$

Suppose we invented a circular machine consisting of a ring of magnets and an 'electrostatic cavity' consisting of an anode and cathode with holes for the beam to pass through as it is accelerated. If the field is static with no time derivative, the right-hand side of the equation must be zero. The integral of the accelerating field for a path threading the cavity and passing round the ring must also be zero. The accelerator does not work. The same is true if we make an 'electrostatic' linac from a chain of such cavities, each with its entry port at the same potential, since in this case we may simply close the path by returning in the field-free region alongside the cavities. There can, therefore, be no repetitive acceleration unless the magnetic field (and inevitably the electric field) varies with time.

In the classical synchrotron or linac, the field oscillates in a resonant cavity and particles enter and leave by holes in the end walls. In such a resonant cavity, energy is continuously exchanged between the electric and magnetic fields, which are entirely contained within its volume. The time-varying fields ensure a finite energy increment at each passage through one or a chain of cavities. The contribution to the acceleration due to each member of the chain adds up. Since the cavities are all independent, they need not be connected in series electrically. There is no build-up in voltage to ground. In principle, any time-varying electromagnetic field could be used, but for simplicity sine waves are used. The equipment, which creates and applies the field to the charged particles, is known as the r.f., and much of its hardware is derived from telecommunications technology.

In Chapter 1 we also met another class of accelerators, which includes induction linacs and the betatron, in which the magnetic field varies with time. These are devices without r.f. cavities where the pulsed time variation of the magnetic field is exploited directly to generate a change in energy along a path without electric terminals or electrodes. Here again the accelerating force must come from a time-varying field.

The reader may also wonder why electromagnetic fields launched in free space from an antenna or laser, are not used for accelerators. There are two difficulties with such schemes. The first is that the electric vector of a light wave is normal to the direction of propagation. The second is that even a very high energy particle will slip behind the accelerating phase of the field of a free-space wave which is travelling at the velocity of light.

This leads us towards a discussion of the configuration of waves that can be used for acceleration, but we will first briefly recall some of the essential theory of electromagnetic waves.

10.1.2 Waves in free space

First we should examine the general properties of waves in free space.

Figure 10.1 shows a plane transverse electric and magnetic (TEM) wave propagating in free space in the x-direction. Its velocity in vacuum is

$$v = c = \frac{1}{\sqrt{\varepsilon_0 \mu_0}},$$

while in a medium of dielectric constant ε_r and magnetic permeability μ_r this becomes

$$v = \frac{1}{\sqrt{\varepsilon_0 \varepsilon_\mathrm{r} \mu_0 \mu_\mathrm{r}}}.$$

The ratio between the fields is

$$\frac{E}{H} = 376.6\sqrt{\frac{\mu_\mathrm{r}}{\varepsilon_\mathrm{r}}} \quad \text{(ohms)}.$$

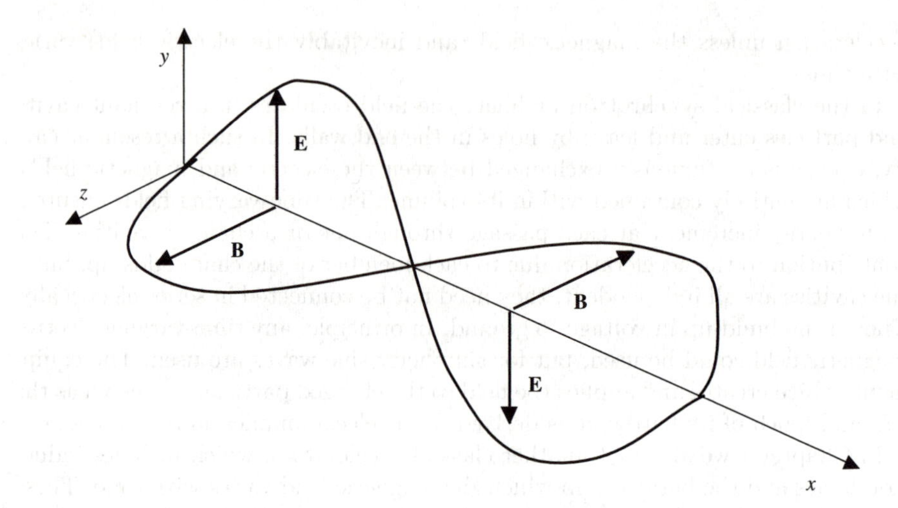

Fig. 10.1 Plane wave in free space.

To be complete, we should recall another useful concept, the Poynting flux or local power flux:

$$P = (\mathbf{E} \times \mathbf{H})\ \mathrm{W\,m^{-2}}.$$

10.1.3 Conducting surfaces

Since, from now on, we will only consider waves in metal boxes, we should recall the boundary conditions of a wave at a perfectly conducting metallic surface. The tangential component of the electric field must vanish and the component of the magnetic field normal to the surface must also vanish. If this were not so, currents would flow in the surface to ensure that it became so. Another quantity associated with the surface is the skin depth. For a wave of frequency f, at a surface of conductivity ρ, this is

$$\delta_{\mathrm{s}} = \frac{1}{\sqrt{\pi f \mu_0 \mu_{\mathrm{r}} \sigma}}$$

and the surface resistance

$$R_{\mathrm{s}} = \frac{1}{\sigma \delta_{\mathrm{s}}}.$$

This will be useful later in calculating the quality of resonators.

10.1.4 Waveguides

Before we try to understand the modes set up in a cavity, let us consider electromagnetic waves in a uniform rectangular waveguide. The propagation along such a guide may be thought of as the superposition of two sets of waves reflected from the sides and interfering as they cross (Huxley 1943).

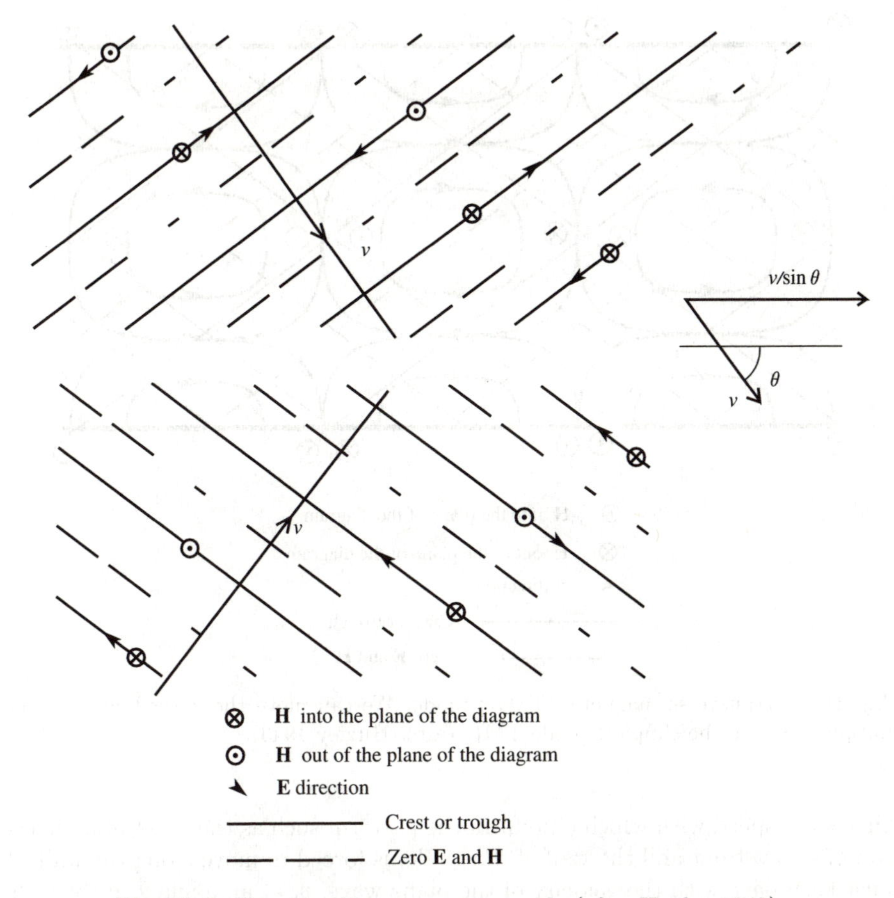

Fig. 10.2 Two travelling waves in a guide (after Huxley 1943).

In Fig. 10.2 we see two plane waves as they might be reflected between two parallel metal boundaries. The lines labelled v indicate their direction of propagation. The solid lines are the crests and troughs of the waves and the dashed lines indicate the zero-crossing. The arrows on the solid lines show the direction of the **E** vector, which lies in the plane of the diagram. The circles with dots and crosses indicate the direction of the **H** vector, which is normal to the plane of the diagram. The dots show **H** pointing towards the reader and the crosses away from the reader.

In Fig. 10.3 we have superimposed these waves to form loops of **E** field sandwiched between the horizontal lines indicating the walls of the guide that reflect the waves. The loops of **E** field enter the conducting walls at right angles as demanded by the boundary conditions for perfect conductivity. Both the component waves of Fig. 10.2 have a velocity from left to right. One might naively expect the velocity of the resulting field patterns along the guide to be $v \sin \theta$, but this would be a mistake. We are interested in the phase velocity of the wave.

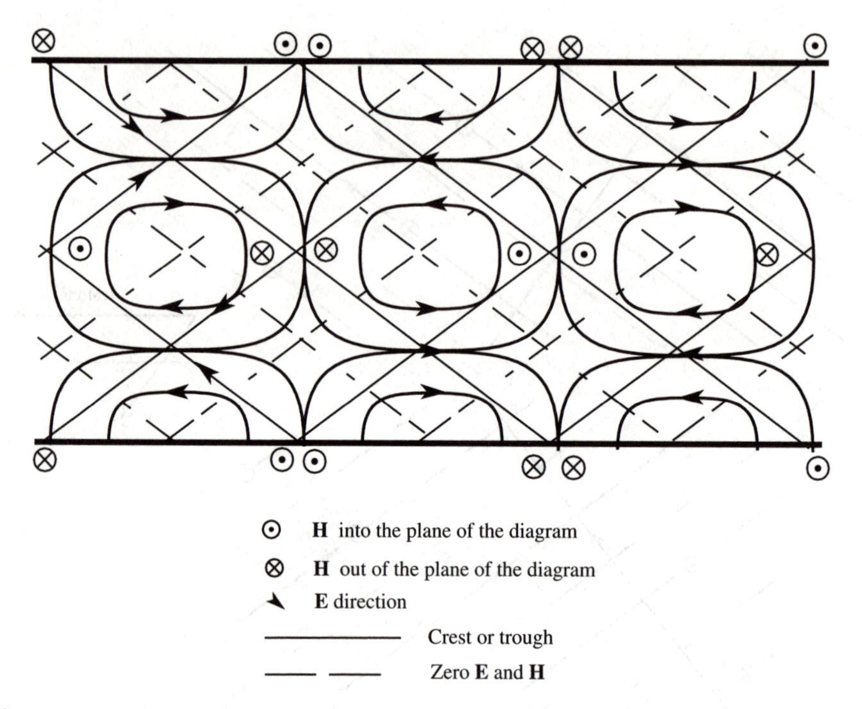

Fig. 10.3 Transverse magnetic (TM_{21}) mode. We can move the lower bound to the mid-plane to see the simplest mode TM_{11} (after Huxley 1943).

This is the speed with which a point in the pattern, such as the point of intersection of a wavefront and the wall of the guide, is forced to move along the wall. It must keep pace with the velocity of the plane wave, v, at an angle θ to the wall. We see in the inset in Fig. 10.2 that the phase velocity along the guide is $v/\sin\theta$ and, therefore, faster than v. If v is the velocity of light, the phase velocity is always greater than that of light. The loops of field must move from left to right with this phase velocity since they are locked to their component waves. This is exactly analogous to the surf boarder moving at a speed higher than that of the waves as he transverses the beach, or a yacht, racing on a broad reach.

Figure 10.3 also shows the direction of **H** which enters and emerges from the diagram forming loops which lie in transverse planes perpendicular to the axis of the guide. The **H** field always points in the transverse direction—hence such modes are called transverse magnetic (TM) modes. There is also a complementary set of modes in which the **E** field is purely transverse but these transverse electric (TE) modes are clearly ruled out for acceleration along the axis of the guide. A more detailed description of these modes can be found in Chao and Tigner(1998).

The two crossing waves and the pattern they form travel faster than light from left to right, yet they must pass an observer with the same frequency that they would in free space. Their wavelengths, λ_g, in the guide must, therefore, be longer than λ, their wavelength in free space. The free-space wavelength, λ, of the two component free-space waves in Fig. 10.3 is less than the width of the guide. If we increase λ, there comes a point where it becomes comparable with the width of the guide. There is a critical wavelength λ_{c} beyond which waves cannot propagate and, below that value, λ_{g} is determined by

$$\frac{1}{\lambda_{\mathrm{g}}^2} = \frac{1}{\lambda^2} - \frac{1}{\lambda_{\mathrm{c}}^2}.$$

It may help the reader to imagine a triangle whose sides are wavenumbers such that $k^2 = k_{\mathrm{g}}^2 + k_{\mathrm{c}}^2$.

Fascinating as this may be, we cannot use waves travelling faster than c for the sustained acceleration of particles which themselves are travelling slower than c. We can see in Fig. 10.3 that the direction of $\mathbf{E}$ alternates along the axis of the guide and particles accelerated by the fields from one loop would be decelerated as the wave overtakes them and they fall into the decelerating fields of the next loop. Nevertheless, let us pursue this discussion of waveguides for there is still much to be learned that can be applied later to cavities.

10.1.5 Group velocity

The fact that the phase velocity of waves can be greater than that of light may seem to contradict special relativity, but we should remember that neither energy nor information is propagated with the *phase* velocity. Energy and information travel with the group velocity of a wave. To understand how group velocity is defined, let us consider Fig. 10.4. Here we see two continuous waves of slightly different frequencies interfering. Their sum can be expressed by the product of two waves:

$$\begin{aligned} E &= E_0 \sin[(k + dk)x - (\omega + d\omega)t] + E_0 \sin[(k - dk)x - (\omega - d\omega)t] \\ &= E_0 \sin[kx - \omega t] \cos[dkx - d\omega t] \\ &= 2E_0 f_1(x,t) f_2(x,t). \end{aligned}$$

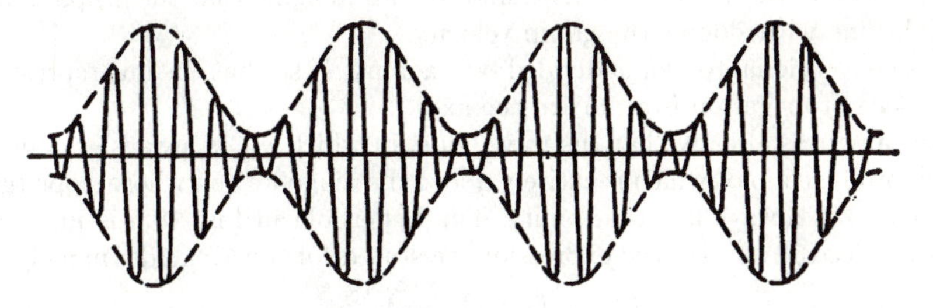

Fig. 10.4 A short wavelength disturbance travels with the phase velocity within an envelope propagating with the group velocity.

The first term in the last line is a continuous wave which has the mean wavenumber of the two waves and their mean frequency. It has the form

$$f_1(x,t) = \sin[kx - \omega t].$$

Any given phase in this wave, such as when it crosses zero, is propagated so that $kx - \omega t$ remains constant. The *phase velocity* of the wave must then be

$$v_\mathrm{p} = -\frac{\partial f(x,t)/\partial t}{\partial f(x,t)/\partial x} = \frac{\omega}{k}.$$

The second term in the product describes the envelope of the pattern in Fig. 10.4:

$$f_2(x,t) = \cos[dk\, x - d\omega\, t].$$

Again, we can argue that any point in the envelope must propagate such that

$$x\, dk - t\, d\omega$$

is unchanged and hence its velocity, the *group velocity*, is

$$v_\mathrm{g} = -\frac{\partial f_2(x,t)/\partial t}{\partial f_2(x,t)/\partial x} = \frac{d\omega}{dk}.$$

In Fig. 10.4 the short-wavelength disturbance moves with the phase velocity while the envelope travels with the group velocity. It is rather easy to see why information, in the form of a meaningful modulation of the carrier can only be transmitted with the group velocity. In order to appreciate why this also applies to energy, it may help the reader to imagine the extreme example of such modulation—an isolated wave packet or delta function in which all the energy is localized. No information or energy can precede or follow its arrival.

10.1.6 Dispersion diagram for a waveguide

We can describe the propagation through a medium, or down a waveguide by plotting a graph of frequency, ω, against wavenumber $k = 2\pi/\lambda$. We can even imagine an experiment in which we inject signals of different frequencies down a pipe and measure the wavelength of the modes transmitted. The ratio ω/k at any point in this plot will give the phase velocity for that frequency or wavenumber and the slope of the tangent, the group velocity. In free space this ratio for each point is just c, the velocity of light, and we can imagine that the graph is just a straight line with slope c the group velocity.

It is conventional to plot ω divided by c against k, so that the line representing free-space propagation is at 45° to the axes.

We can guess already that the waveguide, in which v_ph is always greater than c, will produce a plot which is entirely above the diagonal and whose slope (group velocity/c) is always less than unity. The hyperbola in Fig. 10.5 is just such a curve. Indeed if we take the expression we stated for a waveguide, namely

$$\frac{1}{\lambda_\mathrm{g}^2} = \frac{1}{\lambda^2} - \frac{1}{\lambda_\mathrm{c}^2},$$

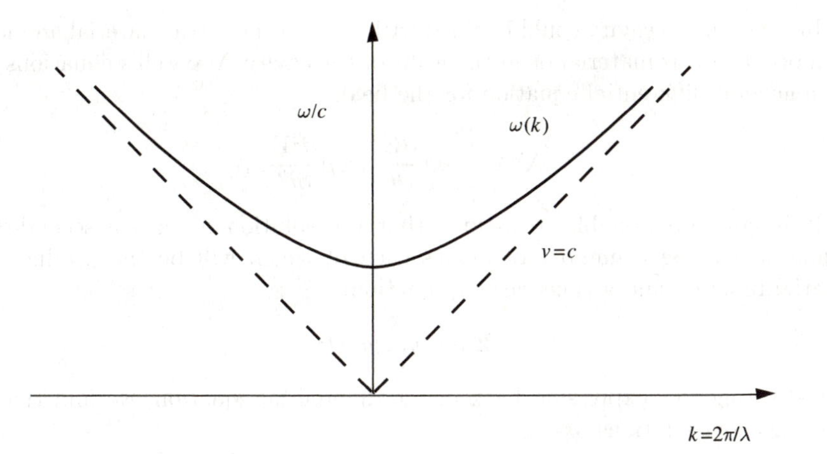

Fig. 10.5 Dispersion diagram for a waveguide.

we can write $2\pi c/\omega = \lambda$ and $2\pi c/\omega_c = \lambda_c$, for the free space and cutoff wavelengths, respectively, and express the wavenumber in the guide as $k = 2\pi/\lambda_g$ to give

$$k^2 = \left(\frac{\omega}{c}\right)^2 - \left(\frac{\omega_c}{c}\right)^2.$$

If we rearrange this,

$$\left(\frac{\omega}{c}\right)^2 = k^2 + \left(\frac{\omega_c}{c}\right)^2,$$

we see that it is clearly the hyperbola in Fig. 10.5.

Note that in this dispersion diagram a small k indicates a very long wavelength but, however small the k, the frequency is always greater than the cutoff frequency. Another interesting point is that the longer the wavelength or lower the frequency the slower is the group velocity and at cutoff frequency no energy flows along the guide. One can show that $v_{ph}v_g = c^2$.

The reader may still wonder why we bother emphasizing these points for a structure which cannot be used to accelerate. We shall soon see that it is possible to find structures which do accelerate and their properties show up as modifications to the dispersion diagram.

10.1.7 Cavity resonators

Waveguide modes are unsuitable for acceleration because the phase of the wave is faster than the particle. However, a simple resonant cavity in which there is a standing wave can be used in a synchrotron provided the revolution frequency is related to the r.f. frequency by an integer h. We now examine the general form of the field in such a cavity.

In general, the cavity could be filled with a linear dielectric material and losses can occur in this material or in the walls of the cavity. Maxwell's equations lead to a general differential equation for the field:

$$\nabla^2\mathbf{E} = \mu\sigma\frac{\partial\mathbf{E}}{\partial t} + \varepsilon\mu\frac{\partial^2\mathbf{E}}{\partial t^2}.$$

It is not unreasonable to assume that the solution of such a second-order equation will be a number of modes each of which will be the product of a spatial function and a time-varying function:

$$\mathbf{E}_n(x, y, z)a_n(t).$$

Substituting this expression back in the differential equation, we find that E_n are the eigen-solutions of

$$\nabla^2\mathbf{E}_n + \Lambda_n^2\mathbf{E}_n = 0,$$

where Λ_n are a set of real parameters related to the resonant frequencies of the modes. Note that the solutions depend on the boundary conditions, and we assume that the losses due to the finite conductivity, σ, of the walls are small enough that we can still assume that the $\boldsymbol{E}$ field is normal to the walls and that no charges flow in the volume considered, that is,

$$\boldsymbol{n}\times\boldsymbol{E} = 0 \quad \text{and} \quad \boldsymbol{n}\cdot\boldsymbol{H} = 0,$$

where $\boldsymbol{n}$ is normal to the wall.

Substitution in the differential equation, with a finite and large σ yields the differential equation for $a_n(t)$:

$$\ddot{a}_n(t) + \frac{\sigma}{\varepsilon}\dot{a}_n(t) + \frac{\Delta_n^2}{\varepsilon\mu}a_n(t) = 0,$$

whose solution is

$$a_n(t) = e^{-t/\tau}\{A_1\cos\Omega_n t + A_z\sin\Omega_n t\}.$$

Here A_1 and A_2 are constants of integration, which depend on the initial condition; Ω_n are the resonant frequencies of the lossy cavity related to the lossless frequency ω_n and the eigenvalue Λ_n:

$$\Omega_n = \frac{\Delta_n}{\sqrt{\varepsilon\mu}}\sqrt{1 - \frac{1}{4}\left[\frac{\sigma}{\varepsilon}\frac{\sqrt{\varepsilon\mu}}{\Delta_n}\right]^2} = \omega_n\sqrt{1 - \frac{1}{4Q_n^2}}.$$

Note also that the decay time τ of the amplitude and the quality factor Q of the lossy cavity are related as follows:

$$\tau = \frac{2\varepsilon}{\sigma} = \frac{2Q}{\omega_n}.$$

10.2 The cylindrical cavity

We now consider the simplest r.f. cavity (see Fig. 10.6), the 'pill-box'. The accelerating modes of this cavity are TM_{0lm}, where the indices refer to the polar coordinates ϕ, r, and z. The modes with no ϕ variation are described as follows:

$$\nabla^2\mathbf{E} + \Lambda^2\mathbf{E} = 0,$$

$$\frac{1}{r}\frac{\partial}{\partial r}\left(r\frac{\partial\mathbf{E}}{\partial r}\right) + \frac{1}{r^2}\frac{\partial\mathbf{E}}{\partial r} + \frac{\partial^2\mathbf{E}}{\partial z^2} + \Lambda^2\mathbf{E} = 0,$$

$$E_z = E_0 J_0\left(\frac{P_{0l}}{r_0}r\right)\cos\frac{m\pi}{h}z,$$

$$E_r = E_0\frac{m\pi}{P_{0l}}\frac{r_0}{h}J_1\left(\frac{P_{0l}}{r_0}r\right)\sin\frac{m\pi}{h}z,$$

$$\Lambda^2_{0lm} = \left(\frac{P_{0l}}{r_0}\right)^2 + \left(\frac{m\pi}{h}\right)^2,$$

where J_0 and J_1 are the Bessel functions of order zero and of order one, respectively, while P_{0l} is the argument, ζ, of the Bessel function, $J_0(\zeta)$, when it crosses zero for the lth time. $J_0(\zeta)$ looks very much like the first quarter wavelength of $\cos\zeta$ but falls to zero at $\zeta = 2.405$ and not $\pi/2$.

Physically, this ensures that the electric field parallel at the wall is zero. The second index, l, indicates the radial variation while the third, m, controls the number of half-wavelengths in the z-direction. It is interesting to observe that if $l = 1$ and $m = 0$ (see Fig. 10.7(a)) then we have the fundamental accelerating mode and the lines of force are straight, without any variation along z, and the resonant frequency does not depend upon the length h of the cavity. Because $P_{01} = 2.405$ we can write the solution of this mode for all

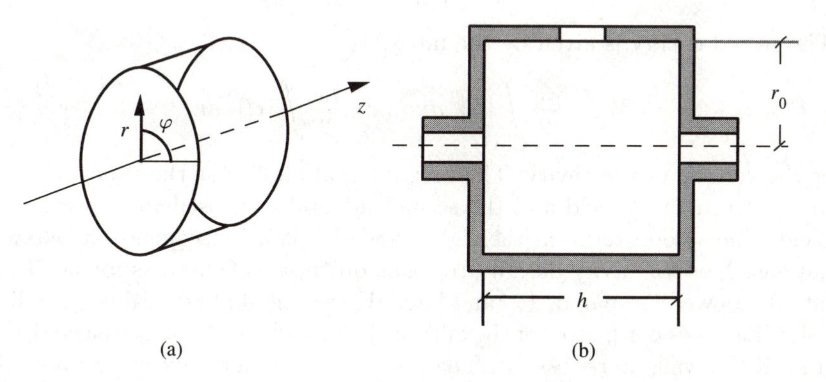

Fig. 10.6 Cylindrical pill-box cavity. Holes for beam and a coupler are shown in (b).

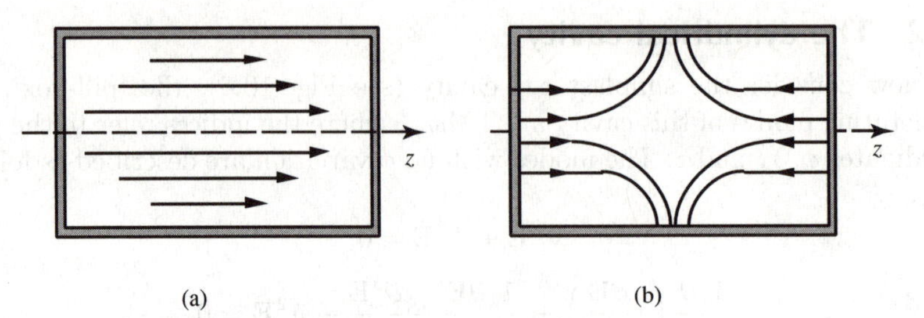

Fig. 10.7 Lines of force for the electrical field as in modes (a) TM_{010} and (b) TM_{011}.

pill-boxes as

$$E = E_0 J_0\left(\frac{2.405}{r_0} r\right); \qquad \Lambda_{010} = \frac{2.405}{r_0}, \quad \omega_{010} = \frac{\Lambda_{010}}{\sqrt{\varepsilon\mu}},$$

$$\nu_{010} = \frac{\omega_{010}}{2\pi} = \frac{1.147\,10^9}{r_0}; \qquad \lambda_{010} = \frac{1}{\nu_{010}\sqrt{\varepsilon\mu}} = \frac{2\pi}{\Lambda_{010}} = 2.61 r_0,$$

and

$$H_\theta = H_0 J_1\left(\frac{2.405}{r_0} r\right).$$

Figure 10.7(b) shows the next mode with one half-wavelength in the z-direction.

10.3 Quality factor of a resonator

The quality factor of a resonator is defined as the ratio of the stored energy, W_s and W_d, the energy dissipated per cycle divided by 2π. The power dissipated is P_d:

$$Q = \frac{W_\mathrm{s}}{W_\mathrm{d}} = \omega \frac{W_\mathrm{s}}{P_\mathrm{d}}.$$

The stored energy is given by the integral

$$W_\mathrm{s} = \frac{\varepsilon_0}{2} \int |E|^2\, dv \quad \text{or} \quad \frac{\mu_0}{2} \int |H|^2\, dv$$

over the volume of the cavity. The first integral applies at the time the energy is all stored in the E field and the second integral as it oscillates back into the H field. The second term on the right-hand side is a consequence of Maxwell's equations. Modern cavity design programs do these calculations for us. To calculate the power dissipated, P_d, and later W_s, we should first evaluate the linear density (amperes per metre) of the current $\mathbf{j}$ along the walls of the selected structure as if the walls were lossless. The losses are then introduced by taking into account the finite conductivity σ of the walls.

Since, for a perfect conductor, we have $\mathbf{j} = \mathbf{n} \times \mathbf{H}$, we can write

$$P_{\mathrm{d}} = \frac{R_{\mathrm{surf}}}{2} \int_{s} |H|^2 \, ds,$$

where s is the inner surface of the structure, $R_{\mathrm{surf}} = \sqrt{\pi f \mu_0 \mu_{\mathrm{r}} / \sigma} = 1/\sigma\delta$ is the surface resistance (for copper $R_{\mathrm{surf}} = 2.61 \times 10^{-7}\sqrt{\omega}$ (ohms)) and δ is the skin depth. Again, the cavity design program can be used to give the surface integral of field.

10.4 Shunt impedance

The commonly used figure of merit for an accelerating cavity is its shunt impedance R_{s}. This is the parameter which relates the accelerating voltage to the power P_{d} which has to be provided to balance the dissipation in the walls. The voltage V can be defined along a path followed by the beam:

$$V = \int_{\text{path}} |E_z(x, y, z)| \, dl.$$

This, of course, is at a fixed instant in time and does not include the transit-time effect discussed below. Shunt impedance is defined, with the usual factor 2 consistent with peak, and not with r.m.s. voltage as follows:

$$R_{\mathrm{s}} = \frac{V^2}{2P_{\mathrm{d}}}.$$

In the world of linear accelerators, the shunt impedance per unit length and power dissipated per unit length are often quoted, frequently without the factor 2.

10.5 The transit-time factor

The accelerating gap may be the space between drift tubes in a linac structure or simply the entrance and exit orifices of a cavity resonator. In calculating the increment of energy given to the particle as it crosses the gap we must take into account that the field is varying as it does so. This makes the cavity less efficient and the particle sees a resultant energy gain which is only a fraction of the peak voltage between the electrodes. This fraction is known as the 'transit-time factor'. Let us now examine it for a simple accelerating gap (Fig. 10.8) supposing that E_z is, at a given instant, uniform along the axis of the gap but depends sinusoidally upon the time:

$$E_z = E_0 \cos(\omega t + \varphi).$$

The phase φ is referred to that of the particle which is in the middle of the gap, $z = 0$, at $t = 0$.

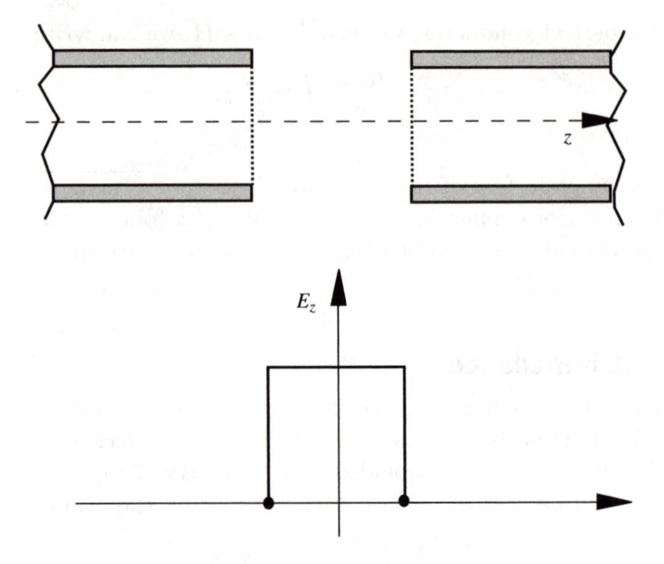

Fig. 10.8 The r.f. gap.

Normally, the energy imparted in a single pass is small compared to the kinetic energy of the particle and we can assume that the speed of the particle does not change during the transit. Consequently $\dot{z} = \beta c$ and the energy gained over the gap, G is

$$V = \int_{-G/2}^{+G/2} E_0 \cos(\omega t + \phi)\, dz = \left(\frac{\sin(\omega G/2\beta c)}{\omega G/2\beta c} \right) E_0 G \cos\phi.$$

The factor in the brackets is known as the transit-time or gap factor and may be already familiar to some readers who have studied optical diffraction. We define a transit angle, in terms of the free-space r.f. wavelength, λ: $\theta = \omega G/\beta c = 2\pi G/\beta\lambda$. The transit-time factor then becomes

$$\Gamma = \frac{\sin\theta/2}{\theta/2}.$$

At relativistic energies the dimensions of a cavity are comparable with half the free-space wavelength and the reduction in efficiency due to the transit-time factor is acceptable. However, at low energy this is not the case and the cavities used often have a strange re-entrant configuration to keep G short compared to the dimensions of its resonant volume.

Increasing the ratio of volume to surface area of the interior of the cavity reduces ohmic losses and improves the Q factor. The shape shown in Fig. 10.9 represents a compromise between this and the need to minimize the gap factor. For obvious reasons the indentations are called 'nose-cones'. Naturally, the gap dimension may also be limited by electrical breakdown.

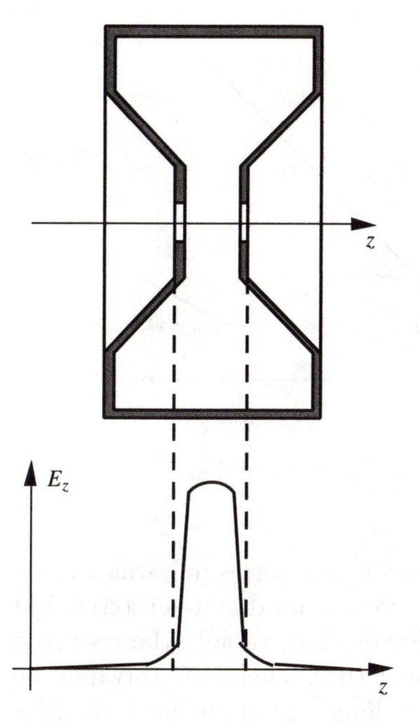

Fig. 10.9 Field in a resonant cavity.

10.6 Iris-loaded structures

Accelerating systems for particles close to the velocity of light often consist of a series of cavities in a single assembly. An example is shown in Fig. 1.6. We may think of these structures as a number of pill-boxes, weakly coupled to each other by the fields which lead through the beam aperture, or alternatively as a cylindrical waveguide, loaded with a number of equidistant irises. Usually, power from an amplifier is coupled into a cavity at one end and is either absorbed in a load at the other end or reflected to set up a standing wave.

The effect of the irises is to slow down the phase velocity of the wave. We saw in Fig. 10.5 that a cylindrical waveguide cannot be used for sustained acceleration because all points in the dispersion curve lie above the diagonal and hence the phase velocity is always greater than that of light. However, an iris-loaded structure has quite a different dispersion diagram in which the function linking the frequency ω to wavenumber k oscillates within a passband of frequencies and replicates like a cosine function at intervals of wavenumber spaced by $2d$ (Fig. 10.10). Usually we are interested only in the first upward slope of this undulating function in the interval $0 < k < \pi/d$ but we see that much of this is below the $v = c$ line, where the phase velocity may be matched to that of the particle by a suitable choice of frequency.

The reason for this dramatic change in the diagram due to the irises is not easy to explain rigorously. An array of oscillating independent pill-boxes can, of course, have any arbitrary phase relation in z, with each other. We would not expect the phase and group velocities of a disturbance travelling from one cell

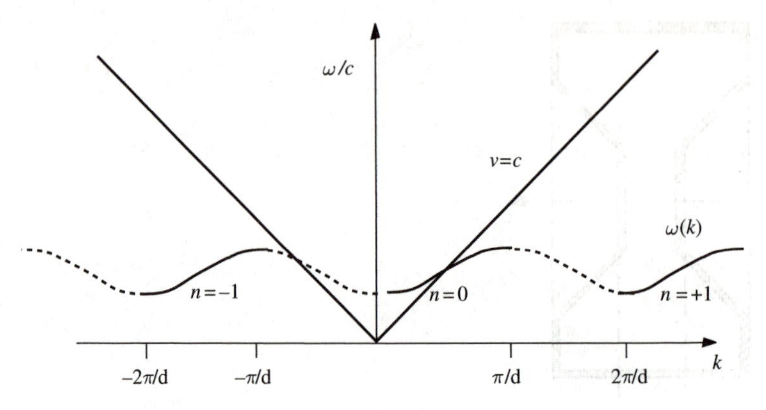

Fig. 10.10 Dispersion diagram for a loaded waveguide.

to the next to be closely related to our arguments about waves in parallel-sided pipes. Waves launched at one end of the structure are no doubt scattered and reflected from the irises and, it is not easy to predict how rapidly the resultant of their interference pattern travels along the structure. There will certainly be reflected paths within the structure that are much longer than the distance along the axis and we might expect this to slow down the wave to ensure that parts of the dispersion curves lie below the line $v = c$.

We may guess that k values of 0 and π/d are standing resonant waves in a single cavity and would, therefore, correspond to the point of zero group velocity and hence to points of inflection in the dispersion diagram. If we expect the waveguide dispersion diagram to be unchanged when $k = 0$ because the field lines are like those in a pipe but distorted to produce a maximum at $k = \pi/d$, it will have to follow the sine-like path of Fig. 10.10. Inevitably, this crosses the line $v = c$ and then the phase velocity becomes less than c.

If the reader is dissatisfied with these general arguments, we must resort to mathematics. Consider an infinitely long chain of such cavities. Symmetry tells us the function that describes their fields $E(r, \theta, z)$ must have the same form in each cavity and can only differ from that in the next cavity by a simple phase factor $e^{-ik_0 z}$. This is called Floquet's theorem and may be written as

$$E(r, \theta, z) = e^{-ik_0 z} f(r, \theta, z),$$

where f is periodic $(f(r, \theta, z + d) = f(r, \theta, z))$ and k_0 is the inverse wavelength of the progression of phase of the wave along the structure and depends on the frequency of the signal injected into the structure.

We can decompose $f(r, \theta, z)$ into a harmonic series of functions which fit the boundary conditions at the walls of a cell and represent a Fourier analysis of the field pattern in a cell

$$f(r, \theta, z) = \sum f_n(r, \theta) e^{-2\pi i n z/d}.$$

The terms of this series are called space harmonics and are equivalent to the higher harmonics of an organ pipe in that n is related to the number of nodes in the cell. We can now substitute the last equation in the one before:

$$\begin{aligned} E(r,\theta,z) &= e^{-ik_0 z}\sum f_n(r,\theta)e^{-2\pi inz/d} \\ &= \sum f_n(r,\theta)e^{-i(k_0+2\pi n/d)z} = \sum f_n(r,\theta)e^{-ik_n z}, \end{aligned}$$

and then by comparing exponents of the last two terms we define a k value for each space harmonic:

$$k_n = k_0 + \frac{2n\pi}{d}.$$

This analysis shows that if we choose any frequency in the dispersion diagram it will intersect the dispersion curve at k values spaced by $2n\pi/d$. The curve must, therefore, be periodic and extend to where the phase velocity is much less than c.

In practice, it is only the first rising slope that is used for acceleration and although the structure will transmit the higher space harmonics and these will dissipate power in the surfaces, they are of little use.

10.7 Synchronizing the particle with cavities

Clearly, if an accelerator has two or more cavities, we want the bunch of particles to arrive at the same phase with respect to the voltage at each cavity. The simplest way to achieve this is to space the cavities by a distance L that a particle travels in one r.f. period

$$L = \beta\lambda,$$

where $\beta = v/c$ and $\lambda = 2\pi c/\omega$, the free-space wavelength of the r.f. excitations. Alvarez achieved this in his linac structure by increasing the distance L between the accelerating gaps along the structure. A snapshot of the fields across each gap would show them all exactly in phase. Earlier, Wideröe had used a different structure in which alternate drift tubes were grounded and a snapshot would show a field vector alternating in sign from gap to gap. The synchronism condition for such a structure is

$$L = \beta\lambda/2.$$

If we have a series of cavities forming an accelerating structure, the particle's r.f. phase advance between cells can be 2π, corresponding to the Alvarez structure, or π, corresponding to Wideröe's configuration.

Figure 10.11 shows snapshots of the fields in two adjacent cavities for (a) π and (b) 2π modes. If we examine the directions of currents in the central partition of (b) we find that they cancel and it is easy to see how, if this partition is omitted, it becomes Alvarez' structure shown in Fig. 10.12.

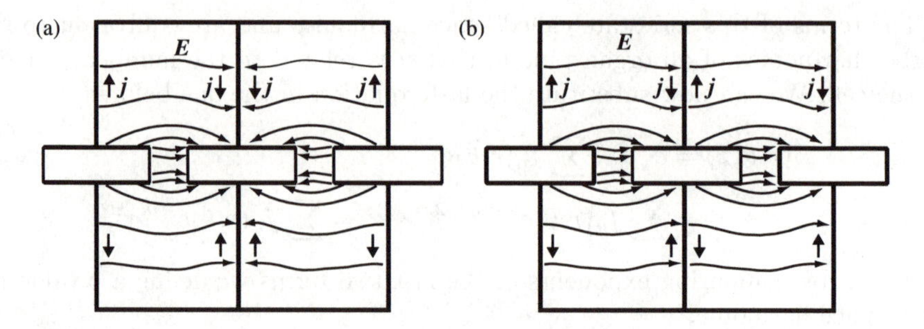

Fig. 10.11 Adjacent single-gap cavities in (a) π mode and (b) 2π zero mode.

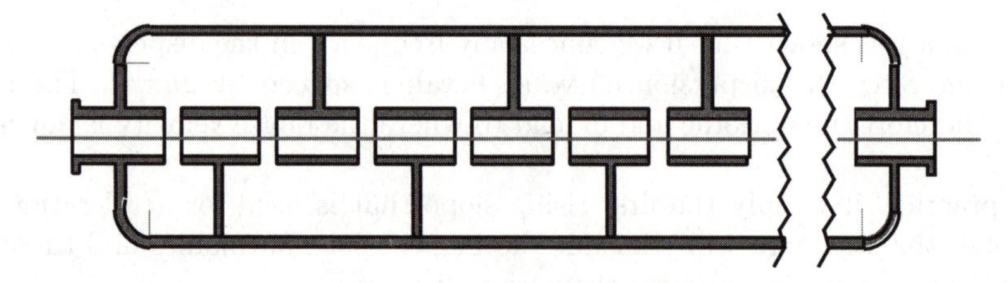

Fig. 10.12 Alvarez cavity.

Neither of these conditions is ideal when the structure consists of a large number of such cavities. The two modes correspond to all the cells oscillating in phase or antiphase and correspond to the ends of the dispersion curve, where $d\omega/dk$ is zero. When the phase advance between cells is such a simple fraction it becomes very difficult for power to propagate along the guide and small errors produce serious distortions. See also Lapostolle and Septier (1970), Wilson (1982), and Loew (1983).

10.8 Other modes of multicell cavities

We are not obliged to use standing-wave configuration and often in an electron linac a wave travels from one end of a long chain of cavities to be absorbed in a load at the other. In such a travelling wave scheme we may choose any point on the dispersion curve. Alternatively we may excite a group of cavities such that the wave pattern repeats every 1, 2, or more cells. The point on the dispersion curve that we select is then linked to the number of cells in which the field pattern repeats. In Fig. 10.13 we see patterns which repeat every one, two, three, and four cavities corresponding to phase changes of 0, π, $2\pi/3$, and $\pi/2$ per cell, respectively. The frequencies accessible on the dispersion band which preserve this phase advance will, respectively, be two, three, four, and five.

For example, if we choose the number of cavities to be a multiple of three rather than two we can have field patterns which repeat every three cavities with a k value of $2\pi/3$, safely on the sloping part of the dispersion curve. Here the group velocity is finite and errors are not amplified. The structure has the

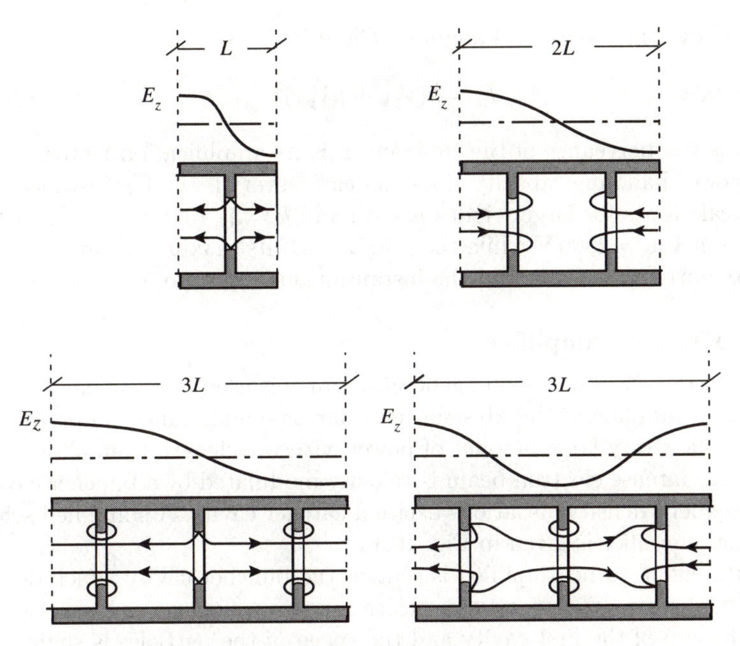

Fig. 10.13 Modes of a multicell cavity.

disadvantage that one of the three cells has a very little effect on the beam but structure designers modify its shape to be still matched in frequency but shorter, or place it to lie alongside the other two cavities in order to economize in length. Such 'side-coupled' structures are favoured at low energies where the distance travelled by the particles per period tends to be short.

The design of acclerating cavities is a vast field and were commend the reader to consult the many standard texts on the subject: Montgomery *et al.* (1948), Moreno (1958), Collin (1961, 1966), Slater (1969), and Ramo *et al.* (1984).

10.9 The r.f. power generation

The sinusoidal power needed to drive the accelerating structures ranges between a few kW and a few MW (continuous wave). The basic elements of most r.f. power amplifiers are the triode or the tetrode with which it is possible to cover a frequency range from a few MHz to a few hundred MHz. At higher frequencies another device is preferred, the klystron. In order to maximize the output power, the input and output circuits are themselves resonant.

10.9.1 Triode amplifier

In a triode, the current I_{a} depends upon the plate and the grid voltages with respect to the cathode. Let V_{pk} and V_{gk} be those voltages. Roughly, the anode current obeys the 'adapted' Langmuir–Child law,

$$I_{\mathrm{a}} = k(V_{\mathrm{pk}} + \mu V_{\mathrm{gk}})^{3/2},$$

where k is the perveance of the tube and μ is its amplification factor.

The power-handling capacity of triodes can be very large. For instance, assume the typical values for large triodes $\mu \approx 40$ and $k = 3 \cdot 10^{-5}$ $(A/V^{1.5})$, then with a minimum $V_{\mathrm{pk}} = 2000\,\mathrm{V}$ while the grid attains its maximum, say $V_{\mathrm{gk}} = 300\,\mathrm{V}$, the plate current is ~50 A and the instantaneous input power is 100 kW.

10.9.2 Klystron amplifier

While the triode is a wide-band generator which is used to make narrow or wide-band tuned amplifiers, the klystron is a narrow-band, tuned amplifier capable of delivering a very large amount of power with wavelength from about 1 m to a few cm. An intense electron beam is velocity modulated by a buncher cavity and the subsequent density variations excite a catcher cavity. A simplified scheme of a klystron amplifier is given in Fig. 10.14.

The r.f. signal to be amplified is sent to the buncher cavity which develops a voltage at the gap. The continuous electron beam which comes from the cathode enters the gap of the first cavity and the speed of the particles is slightly varied according to the phase of the voltage at the entrance. In this way, averaged over more than one cycle, the uniform beam comes out slightly modulated in velocity. This operation does not involve an energy exchange between the cavity and the

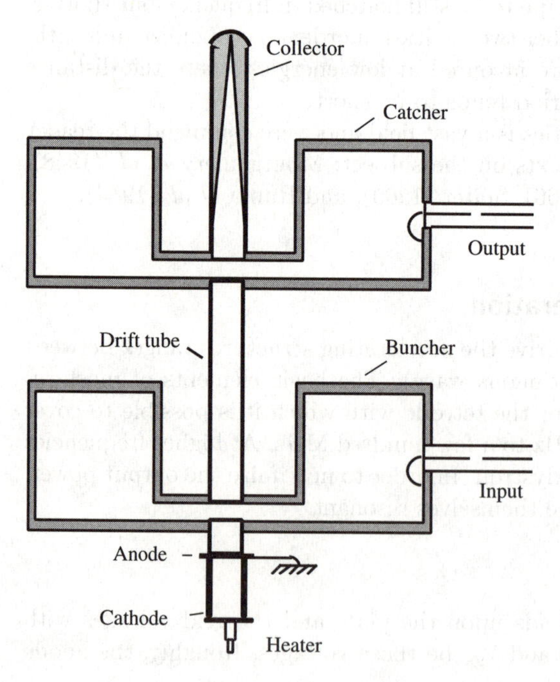

Fig. 10.14 Schematic of a klystron amplifier.

beam as long as the entering beam is uniform, and the changes in the speed are small. The emergent beam travels along the drift tubes and, due to the differences in the speed of the particles, undergoes the process of bunching. (This is a very complicated process, especially if the space-charge effects are taken into account.)

At some distance from the buncher, there is a particular position, the first focus, where the electrons arrive in bunches which, theoretically, have infinite longitudinal density. The gap of the second cavity, the catcher, is located at this point and becomes excited by the train of bunches arriving at the r.f. frequency. Output power is excited in the cavity and absorbed by a coupling loop. In this process, part of the kinetic energy of the electrons coming from the cathode is converted into r.f. power. Having lost the greatest part of their kinetic energy, the bunches inside the catcher are finally absorbed by the water-cooled collector.

Often two or three 'idle' cavities are inserted between the buncher and the catcher as refinements to this simple design which improve the bunching action. A focusing solenoid is placed along the bunching region.

The pulsed power from an industrial klystron can be as large as 50 MW. In continuous-wave operation a power of over 1 MW has been reached at 350 MHz.

10.10 Coupling

It is beyond the scope of this book to discuss the many ways that power may be coupled to the cavities, but we show two examples in Fig. 10.15 in which the electrical power excites a loop that is coupled to the cavity. This means that the magnetic field created by the loop should have a component in common with

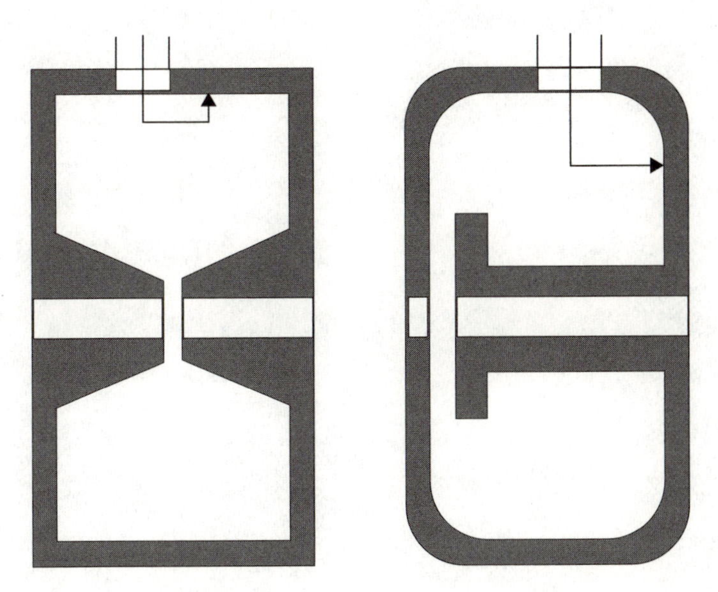

Fig. 10.15 Two examples of loop coupling.

the magnetic field of the mode we wish to excite in the cavity. As shown in Fig. 10.15, the loops are placed in the region of the cavity where the magnetic field is stronger.

Exercises

10.1 What is the fundamental resonant frequency of a pill-box cavity of length 50 cm and diameter 75 cm?

10.2 What is the resonant frequency of the next highest (TM_{011}) mode?

10.3 What kinetic energy of protons would be required to ensure a transit-time factor of 41% for the fundamental mode of this cavity?

10.4 How would you modify the accelerating system to accelerate lower-energy protons?

10.5 Assume a simple pill-box cavity. Use the expressions in Section 10.3 to show that the quality factor is related to the surface S, volume V, and skin depth

$$\frac{2}{\delta} K \frac{V}{S},$$

where K is a form factor.

10.6 Using Fig. 10.10 and assuming a three-cell cavity, plot the points on the first, $n = 0$, arm of the diagram which correspond to the appropriate k value, estimate the phase and group velocities for these points by inspection.

11 Colliders

11.1 Twin-ring colliders

In Chapter 1 we outlined the early history of colliders. Suggestions from the MURA Group (Kerst *et al.* 1956) led immediately to the Stanford–Princeton proposal for electron–electron collider rings, completed only in 1966 (Barber *et al.* 1966), and the VEPP-I electron–electron collider at Novosibirsk, started in 1962 (Budker 1964) and completed in 1965. These were all twin-ring schemes in which collisions take place when the beams of two synchrotron rings, one circulating clockwise, the other anticlockwise, meet at a crossing point common to both rings. This was the configuration adopted later for the first large proton–proton ring, the ISR (Johnsen 1964) completed in 1971 at CERN, and will also be that for the LHC.

To achieve useful 'event rates', the beam current densities have to be very high. To this end, the builders of these early machines and their colleagues at the Cambridge (Mass.) Electron Accelerator strove to accelerate and store the high-beam currents. They made many notable contributions to our understanding of beam behaviour. The designers of the ISR continued in this tradition. In the ISR the two 30 GeV rings were interlaced to cross at eight points around the circumference. The ISR vacuum system had to be extremely good to avoid phenomena in which the beam was adversely affected by residual ionized gas, but even then there was more to be learnt when the ISR was switched on. A 'brick wall', due to local cancellation by space change of the chromaticity needed to stabilise the beam, as well as the effect of ionisation in the vacuum chamber had both to be understood. To achieve sufficient event rates, each beam had to contain many amperes of current circulating without the help of r.f. buckets, which would have only made the beam unstable. Small r.f. buckets were used to nudge the beam by 'phase displacement' from its injection orbit to join the accumulated stack of protons. Many of the ISR's novel features later became standard practice in other proton machines.

11.2 Single-ring colliders

These twin rings were not however the first colliders. The first collider ring (Amaldi 1981) was built by Touchek. This predated the Stanford and VEPP-1 collider and consisted of a single ring. The oppositely charged electrons and positrons circulated in opposite directions in the same guide field and, being equal in mass but of opposite charge appeared to the magnetic bending and focusing fields as identical currents, one bent to the left, the other to the right.

This idea was subsequently exploited in a number of larger colliders including PEP at SLAC (Paterson 1980), PETRA at DESY (Degèle 1980) and, of course, LEP at CERN.

11.3 Proton–antiproton colliders

The single ring idea can be applied to protons and antiprotons, however antiprotons are much more difficult to produce than positrons and building the ISR as a single-ring proton–antiproton machine would not have been technically feasible at the time.

It takes a pulse of 10^{13} protons to produce 10^{6} antiprotons and it takes tens of thousands of pulses of protons to produce enough antiprotons for a proton–antiproton collider. Liouville's theorem dictates that, if several antiproton pulses are injected into an accumulator ring, they can only be placed side by side rather than be superimposed. The ring would soon be full and the density of the antiprotons be much too diffuse to produce a reasonable probability of encounter in a collider. The discovery of a method of phase-space compression, or cooling, offered a way round this conservation theorem by reducing the longitudinal and transverse oscillations of individual particles in the manner of Maxwell's Demon. Electron cooling was invented by Budker (1967) at Novosibirsk, and stochastic cooling by van der Meer at CERN in 1968, to be (published much later; van der Meer 1972). These methods made it possible for antiproton beams to be concentrated and stacked in the vacuum chamber of the ring, thus accumulating antiprotons harvested over many hours of production.

The first antiproton accumulator was built at CERN and the 400 GeV SPS pulsed synchrotron was converted into a proton–antiproton collider (Evans 1984). This was followed several years later by Fermilab near Chicago, which built a superconducting ring of 1 TeV (1000 GeV) nominal energy, the Tevatron, which still operates as a proton–antiproton collider (Griffin 1980).

11.4 Electrons versus protons

Protons are complex objects consisting of three quarks held together by gluons. In collision with other hadrons—protons or antiprotons—the interaction is dominated by this strong force. However, only one quark in each of the colliding hadrons is involved in the interaction and it is difficult or impossible to identify which quarks or gluons have taken part in any given interaction. In addition, the quarks which interact carry only about 10% of the total energy given to the composite particles by the accelerator, so that a large fraction of the energy is wasted. Nevertheless, hadrons are unique probes for the study of the strong interaction.

Electrons, on the other hand, are ideal probes for electromagnetic interactions but unable to probe differences between quark flavours. On the other hand, the point-like nature of the electron ensures that all its energy is put to good use in producing particles of interest.

11.5 Recent circular colliders

It was principally this argument that led CERN to choose a 100 GeV electron–positron machine LEP (CERN 1984) as the collider to follow the SPS. This has four times the diameter of the SPS collider but only one-third of its energy per beam. The reason for the drop in energy is that electrons, 2000 times less massive than protons, have velocities closer to the speed of light than protons of the same energy. Any charge circulating in an accelerator will emit electromagnetic radiation but, when its velocity approaches that of light, the loss of energy by the particle multiplied by the number of particles in the colliding beams rises steeply. This represents a large power, often many MW, which must be replaced from the electric supply via the r.f. cavities. This can be reduced by bending with a large and gentle radius—hence the size of LEP. The whole topic of this synchrotron radiation, as it is called, was discussed in Chapter 8.

While CERN constructed LEP, DESY in Hamburg built two rings to collide 30 GeV electrons with 820 GeV protons which first operated in 1990 (HERA 1981). HERA was the first European machine to use superconducting accelerator magnet technology for the whole ring and is unique in its ability to study the physics of hadron–lepton collisions (DESY 1981). The next step in the quest for higher energies, this time to produce the massive Higgs particle, is the construction of LHC, a 7 TeV on 7 TeV separate ring collider.

The magnet cross section of the 7 TeV hadron collider the LHC (Lefèvre 1995), is shown in Fig. 11.1. It makes use of CERN's existing injector synchrotrons and is to be installed in the existing LEP tunnel. At the end of 1994, the CERN Council agreed to a ten-year programme to construct the machine. This may be the last circular collider to be built, but hopefully not the 'last great machine'.

A potential problem in LHC and beyond is that at these high intensities and energies even proton beams will radiate a significant flux of synchrotron light. The power that loads the refrigerator for the whole ring is

$$P_{\text{syn}} = \left(\frac{4\pi}{3}\right) N f_{\text{b}} r_{\text{p}} m_{\text{p}} c^2 \left(\frac{\gamma^4}{\rho}\right).$$

This power is expensive to remove at superconducting temperatures.

11.6 Limits and how to overcome them

One of the challenges of experiments with any collider is the need to increase the probability of collision, or luminosity. This is necessary in order to keep abreast of the production rate of the interactions of interest to high-energy physicists.

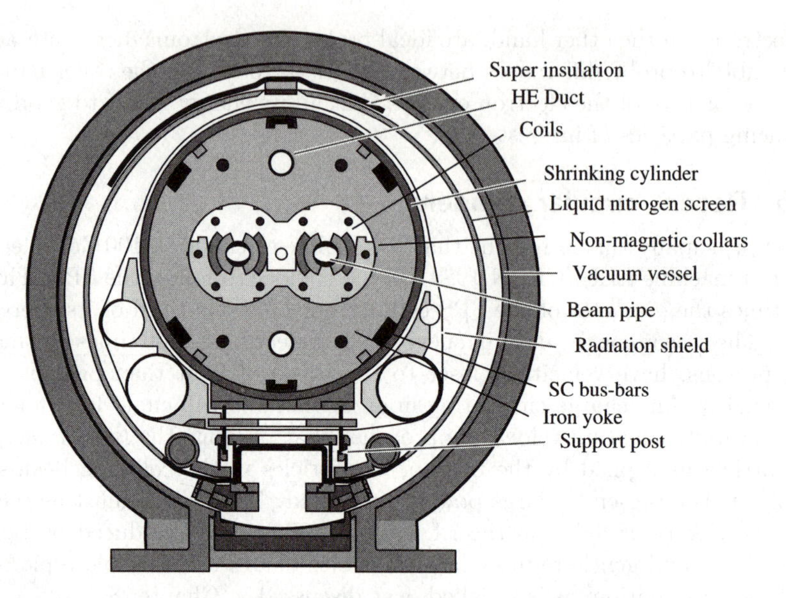

Fig. 11.1 Cross section of the LHC twin-bore dipole magnet in its cryostat.

Such cross sections fall as $1/M^2$, where M is the mass, see line labelled σ_{jet} in Fig. 11.2 from Amaldi (1987). Likening the accelerator to a microscope using particles as de Broglie waves of short wavelength, $\lambda = h/p$, to discern fine details of structure, a higher energy produces a shorter wavelength to reveal smaller objects but these smaller objects scatter less.

11.6.1 Luminosity

Let us define luminosity precisely. Take a probe particle in one beam which sees the oncoming beam as a cloud of N particles, each a disc of cross section σ (Fig. 11.3). The probability of the rare process with cross section σ occurring is just the fraction of the beam's area A that is occluded:

$$P = \frac{N}{A}\sigma = l\sigma.$$

In this trivial case of a single encounter of one particle passing through a beam, $l = N/A$ is a sort of luminosity for this one encounter. It is independent of the cross section under study and depends only on the beam geometry. We may think of it as a probability of producing an event normalized to unit interaction cross section.

In practice, the probe beam has as many particles as its opposing target, and such encounters occur as often as the many bunches in the circulating beams in a storage ring meet each other. The luminosity for two such equal beams colliding

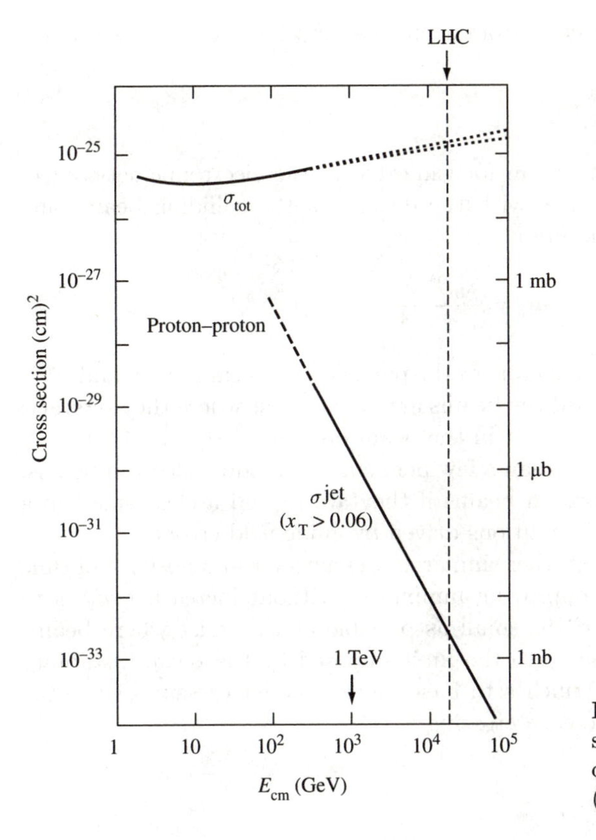

Fig. 11.2 Proton–proton cross sections as a function of centre-of-mass energy from Amaldi (1987).

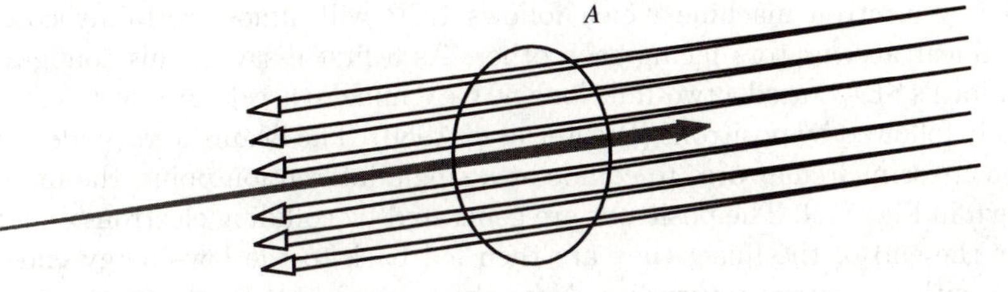

Fig. 11.3 A probe particle encounters a target—a beam of particles with cross sectional area A travelling in the opposite sense.

head on is

$$L = \frac{N^2 f_b}{A},$$

where N is the number of particles per bunch, f_b is the bunch frequency, and A is the transverse beam area at the crossing. A luminosity of $10^{33}\,\mathrm{cm}^{-2}\,\mathrm{s}^{-1}$

will produce one event per second from a process for which the cross section is $10^{-33}\,\text{cm}^{-2}$.

11.6.2 Tune shift

The beam–beam effect, a change in focusing due to the electromagnetic effect of the oncoming beam, increases with the density of the colliding beams and imposes an upper limit on luminosity given by

$$dQ = \frac{r_0\beta^*}{\gamma}\frac{N}{A},$$

where γ is the usual relativistic factor, r_0 the particle's classical radius, and β^* is a measure of how tightly focused the beams are at the waist where they interact: a small β^* represents a narrow waist in the beam profile.

This Q shift must be no more than a few per cent for colliding electron beams, and a factor 10 smaller for proton beams if they are to find a clear space in a forest of potentially resonant conditions driven by guide field errors.

One can see from the above rather similar expressions for luminosity and tune shift that the only route to improving luminosity without increasing dQ is to squeeze the beams to make β^* as small as possible at the waist where beams collide. One can also increase N to the limit imposed by the beam instability and increase the number of bunches to have more frequent crossings since the above considerations apply to each crossing.

11.7 Linear colliders

LEP has stretched the concept of the circular electron collider to the limit, and any electron machine which follows LEP will almost certainly consist of two linear accelerators facing each other. As a first essay in this configuration, Stanford's SLAC used a two-mile long 50 GeV linac to accelerate electron bunches closely followed by positrons (Richter *et al.* 1980). The beams were guided around opposite semicircular arcs to collide at a single interaction point, the final focus shown in Fig. 11.4. The positrons are generated by colliding electrons on a target near the end of the linac; they are then led back to the low-energy end of the linac with a positron return line. Near the beginning of the linac, there are two damping rings; one of these is for electrons and the other for positrons. As the particles circulate in these rings, they radiate some of their energy as synchrotron light in such a way that their emittance shrinks in all three dimensions. This shrinkage is essential in order to make the colliding beams dense enough to give the luminosity needed when they collide.

At the SLAC Linear Collider (SLC) energy of 50 GeV, it is economical to use the arcs to bring the two beams into collision, rather than building a second linac, but for a machine of several hundred GeV, the arcs would radiate too much energy. In choosing parameters for a new machine, most designers make the frequency and repetition rate of the linac much higher than at the SLC and aim

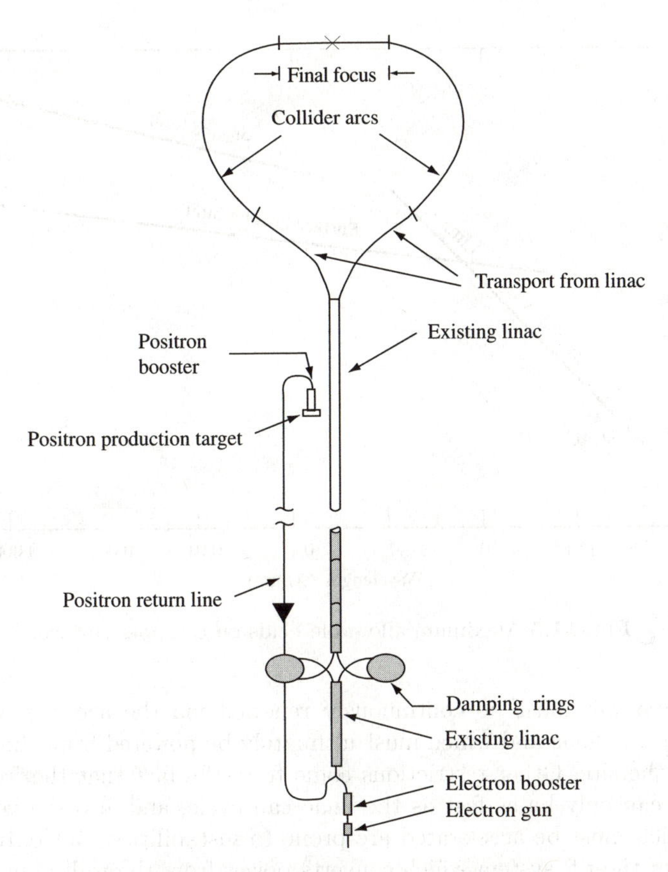

Fig. 11.4 The SLAC Linear Collider.

for a luminosity that can compete with circular machines. The superconducting TESLA project centered at DESY in the exception. The SLC demonstrated that linear colliders are a practical way ahead for lepton physics. The mastery of this new technique took longer than expected and, although SLC cannot match LEP's luminosity, its highly polarized leptons add another dimension to its experiments.

The way forward for leptons has to be a linear collider in which electrons from one linac collide with positrons from another, or perhaps a muon collider, in which it is a race against time to collect, cool, accelerate, and store enough muons for a ring collider before they decay and are lost.

There is an essential difference between linear and circular colliders. Particles once accelerated to high energy in a circular collider may be re-used almost indefinitely, producing a new encounter each time they circulate, and the beam power—the product of the number of particles, the charge they carry, and the voltage to which they are accelerated—is not wasted. On the other hand, in a linear collider a fresh batch of particles must be accelerated for each encounter.

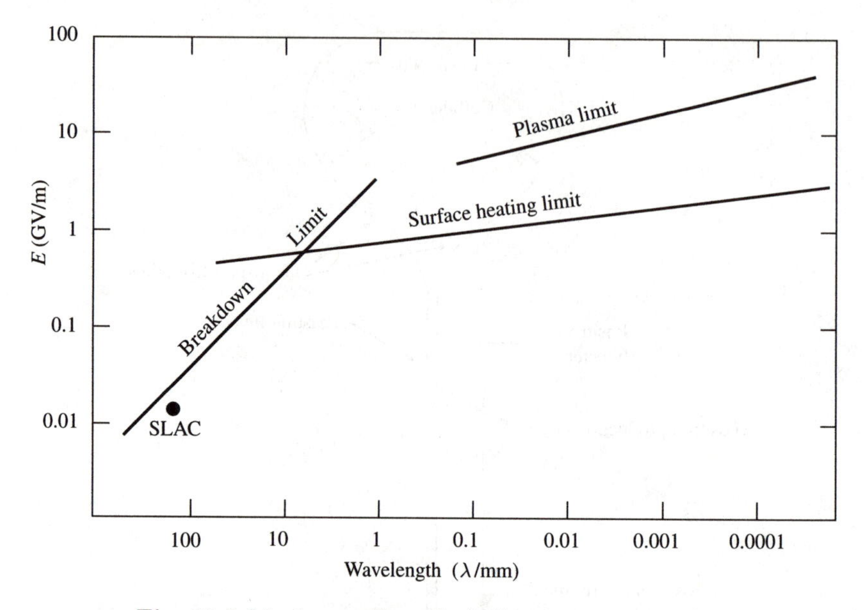

Fig. 11.5 Maximum allowable fields on a copper surface.

The beam power must be continuously renewed via the accelerating cavities which form the linac and which must ultimately be powered from the electricity supply to the site. Other restrictions come from the fact that the frequency of encounter can only be as fast as the linac can cycle, and also the very intense beams which must be accelerated are prone to instabilities. Not only must the efficiency of the r.f. system which converts power from the wall plug into beam power be high, but the voltage gradients in the accelerating cavities must be exceptional if the linacs are to be only 10 km or so in length. Higher-frequency linac structures help, since the energy stored in their fields is smaller and they can be run at higher accelerating field so that the total length of the linac may be shorter. If we proceed in this direction, we have to miniaturize the accelerating cavities to run them at a higher-voltage gradient. Figure 11.5 shows how almost two orders of magnitude in GV/m are to be gained by reducing the r.f. wavelength, which fixes the scale of the cavity dimensions (Weiland 1985). In the next generation of linear colliders, TESLA excepted, the wavelength will be reduced from the 10 cm of present linear accelerators to even 1 cm. However, we must expect instability problems to arise because the walls are closer to the beam. For the highest frequencies it will also be necessary to develop r.f. power sources at frequencies far beyond the bands where power tubes for normal telecommunication use are available. CERN, as part of a world-wide collaboration, is studying the design of a linear collider of at least 2×500 GeV, the minimum required for a significant step into the future (Delahaye 1999).

It is not sufficient just to design an accelerating structure with maximum MeV/m. The beam should sweep up as much as possible of the electrical energy stored in the cavity or the residual energy will be wasted on warming the copper walls of the cavity. If the cavity energy stored per metre of cavity is W' and if N electrons of charge e are accelerated by a field E when passing through the cavity, the sweeping efficiency is

$$\eta = \frac{eNE}{W'}.$$

The r.f. power needed is the power of the beam divided by the sweeping efficiency:

$$P_{\mathrm{rf}} = \frac{P_{\mathrm{beam}}}{\eta} = \frac{NfeU}{\eta}.$$

Here U is the overall energy gain of the particles and f is the repetition rate of the device. We can imagine that small high-frequency cavities are likely to be a good idea because the number of joules, W, needed to fill the interior of the cavity is minimized. In addition, there are the arguments for high frequency based on reducing the overall length of the device. If we try to sweep too much energy from the cavity, the sagging of energy loss between the head and tail of the bunch will be large but this may be compensated to some extent by splitting the bunch into a train of bunches and filling the cavity with energy as they pass.

The parameters f and N in the formula for power appear again in the expression for the luminosity:

$$L = H\frac{fN^2}{4\pi\sigma_x\sigma_y}.$$

Here H is the enhancement of the luminosity due to the electromagnetic forces in the collision (it is unity at low intensity), an effect which is beyond the scope of this description. Comparing the equations for efficiency and power, it looks as if we are free to reach any luminosity by increasing N and reducing f in proportion to keep a constant r.f. power. However, there is a limit to this, because not only does L depend on N^2 but also on the energy spread in the electrons as they radiate photons upon colliding with the bunch of positrons coming the other way. This *beamstrahlung*, as it is called, is caused by the acceleration due to the electromagnetic field of the oncoming bunch. If the fractional energy spread is greater than 20–30%, it will be impossible to define the energy of the incoming particles with the precision needed to interpret the event.

Teams in the USA, CERN, DESY, and Japan have arrived at essentially similar parameter sets for a collider in the range of 0.5 to 1 TeV per beam. The differences stem from a choice in frequency or, in the case of TESLA at DESY, in the use of superconducting cavities. To reach high frequencies, the r.f. power source is the crucial question. You cannot buy a tube that will power a 30 GHz linac off the shelf or even design one. In CERN's CLIC (Delahaye *et al.* 1999), the main linac receives its power from an intense low-energy driver linac operating at the LEP r.f. frequency at which klystrons are available. The drive linac beam consists of very short and intense electron bunches. They must be only a few millimetres long, for they transfer power to the main linac by virtue of their

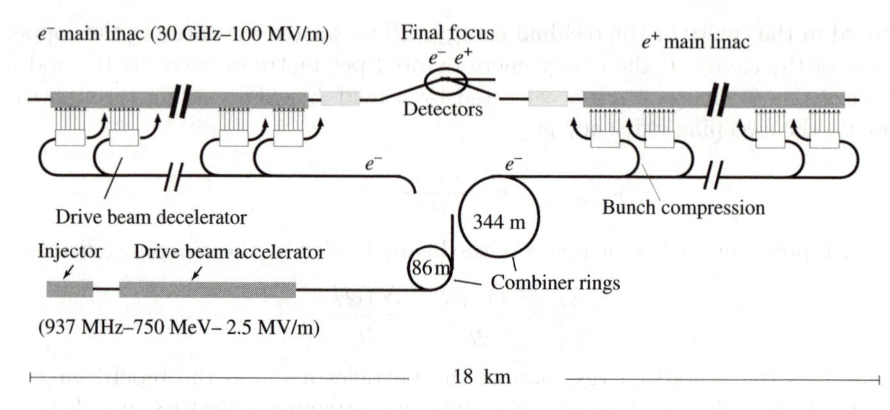

Fig. 11.6 Layout of CLIC at 1 TeV centre-of-mass energy.

strong Fourier component at 30 GHz, which excites waves in a set of transfer structures bridging the two linacs. These may be seen in Fig. 11.6, which also shows the general layout.

11.8 Muon colliders

Recently, there has been a growing realization that muons might well be worthy competitors to electrons. They have the same point-like property as electrons in collision, yet are much further from any synchrotron radiation limitation because of their higher mass. The short lifetime of muons at the energies we might wish to store them has led us to dismiss them as projectiles in collision. However, at higher energy they begin to become interesting if we can provide a copious supply of muons and refill the collider much more frequently. Above a few hundred GeV the Lorentz transformation of their decay is in our favour and one may contemplate a collision time of tens of milliseconds—perhaps long enough for a replenishment to be prepared in a series of rapid cycling accelerators.

There is still a race against time to collect and accelerate the muons. In the low-energy regime, their decay length,

$$\lambda_\mu\,(\mathrm{m}) = 6233p\ (\mathrm{GeV/c}),$$

is short compared to the distance they have to travel to be accelerated. In a synchrotron in which the beam circulates for many turns and is accelerated by a relatively weak r.f. system, it takes too long to reach a given energy. At the opposite extreme lies a plain muon linac but, although this is the most economic in terms of the path length, it brings with it the problems of the linear collider. It seems that the compromise of a recirculating linac is just rapid enough to replenish the muons for good luminosity and is the best solution, limited only by the scale of existing rings and laboratory sites.

The design energy of such a collider might well be as high as 4 TeV and, to be useful, we must aim for a correspondingly high luminosity. As in all colliders, the luminosity is strongly dependent on the beam–beam tune shift that may be tolerated. The beam–beam tune shift limit for electron machines is about an order of magnitude less severe than the proton machine limit because non-linear effects, which are too slow to grow in the damping time of a few milliseconds, will not appear. The muon collider, in which the beam circulates only for a few tens of milliseconds, will be just as forgiving. Muon colliders will not suffer from the energy spread due to *beamstrahlung* and, muons are more potent projectiles for Higgs physics (Fig. 11.7).

The methods of production, collection, and cooling of muons have much in common with the design of an antiproton source, yet there are some very important differences. The final crop of antiprotons, from primary protons hitting a target, yields only one antiproton per million protons. In the case of the muon collider, this is closer to one-to-one. The antiprotons, collected as a 24 GeV proton beam hits a target, have a production peak at a momentum of several GeV while the pions, parents of the muons, are almost all below 200 MeV. Low-energy pions may be produced at a mercury jet or moving belt target. The proton beam, focused on the target, has a mean beam power of several MW, and is produced by a fast cycling synchrotron of typically 16 GeV beam power, similar to those proposed for spallation sources. The low energy of the pions makes the design of a magnetic focusing device to turn the production cone into a forward collimated beam much easier than it was for antiprotons.

The pions are collected in superconducting solenoids with fields tapering from 20 T around the target down to 1 T in the decay channel. A paraxial particle entering a solenoid at a distance r_0 receives an azimuthal momentum,

$$p_\phi = \frac{eBr_0}{2},$$

as it passes through the fringe field. It therefore executes a spiral path whose radius of curvature in the transverse plane is $R = r_0/2$, bringing the particle onto the axis of the solenoid after one half turn of the spiral. Short solenoids of appropriate length, therefore, act as lenses which focus in both planes, and from these one can construct a FOFO lattice with

$$\beta_\perp = \frac{2p_z}{eB},$$

which happens to be numerically equal to twice the radius of curvature of the particle path in the field B.

11.8.1 Monochromator

Now follows the monochromator, which collects a large fraction of the pions as they decay in flight to muons. This consists of a system of r.f. cavities which decelerate the particles that arrive early and accelerate the latecomers. At the same time, it serves as a linac to accelerate and increase the decay length of the muons.

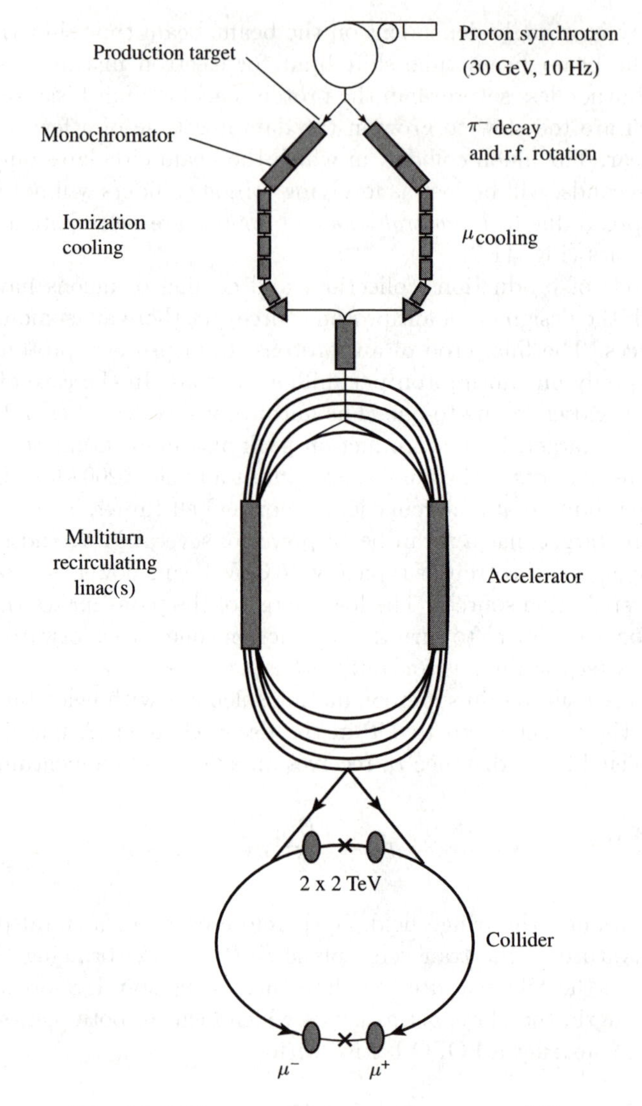

Fig. 11.7 The overall scheme of a muon collider.

11.8.2 Cooling

In Chapter 12 we describe how muons may be cooled by a combination of ionization cooling and emittance exchange and these methods, developed specifically for the muon collider, are currently being studied intensively. Even more exotic optical cooling methods are currently under study.

11.8.3 Recirculating linacs and the collider ring

Recirculating linacs provide the most effective method to bring the muons to high energy. They pass several times through the same linac structure, returning each time through one of several bending arcs matched to their momentum. They finally enter the collider ring itself.

Exercises

11.1 Calculate the luminosity for a beam of 10^{11} protons per bunch circulating at 7.7 TeV in a ring of 26.658 km circumference, assuming $\varepsilon(\beta\gamma) = 3.75$ mm mrad and $\beta^* = 0.5$ m. Emittance is defined for 1σ and there are 4725 bunches per beam. (Assume $A = 4\pi\sigma^2$.)

11.2 Calculate the beam–beam tune shift per crossing for the above.

11.3 Assuming that ΔQ_{bb} should be <0.00083, adjust the number of bunches and intensity to maximize the luminosity within this limit. Assume $N f_{\mathrm{b}}$ remains constant.

11.4 A linear collider seeks to achieve a centre-of-mass energy of 1000 GeV. Plot a curve of length versus field gradient and use Fig. 11.5 to decide the frequency which fits a site of 25 km extent (assume a filling factor of 70%).

11.5 Assuming a repetition frequency of 200 Hz and a mean beam radius ($\sigma_x = \sigma_y = 60$ nm), what beam intensity does the linear collider require to reach luminosity of $10^{34}\,\mathrm{cm^{-2}\,s^{-1}}$?

11.6 What is the average beam power?

11.7 Calculate the lifetime of a muon (a) at 50 GeV circulating in a storage ring and (b) at 4 TeV. Calculate the leading parameters (bending radius Q, number of periods, etc.) for a 6 T superconducting ring to store the muons.

11.8 What would be a reasonable repetition (filling) rate for a muon collider at 4 TeV, assuming that a beam should be renewed when it has decayed by one exponential lifetime?

12 Cooling

12.1 The need for cooling in antiproton machines

In the early 1980s CERN, and later Fermilab, built and brought into operation a completely novel kind of storage ring designed to accumulate and concentrate antiprotons. The SPS and Fermilab proton synchrotrons had originally been constructed to accelerate protons, to extract them and then collide them with fixed targets. It was argued that, if one could collect enough antiprotons, they might be injected into a storage ring and brought into collision with a proton beam circulating in the opposite direction. The opposite charges of the particles and antiparticles (as in electron–positron colliders) would ensure that the opposing beams would see identical bending and focusing forces from the magnets which provided the guide field. For hadrons at these energies (300–900 GeV), the increase in available energy in the centre-of-mass frame would be 20–30 times greater than in fixed-target collisions. It was argued that this might produce enough energy in the centre-of-mass frame to create W and Z particles, whose existence would verify the latest electroweak theory, unifying the laws of electromagnetism and radioactivity—a discovery as important as Maxwell's unification of electricity and magnetism was in the last century.

Antiprotons have always been a rare commodity. The threshold energy for producing them by bombarding a fixed target with protons is just over 6 GeV and protons accelerated to this energy in the Bevatron produced the first trickle of antiprotons in the 1950s. More powerful accelerators, the AGS and PS, followed but, even with primary protons of 25 GeV and above available, about a 1000 protons are needed to produce each antiproton. Moreover, of those produced, only about one in a thousand falls in the narrow range of angles and momenta which can be collected and transported by a beam line to an experiment. Nevertheless, in the 1960s and 1970s, many successful fixed-target experiments were carried out with the pulses of 10^6 antiprotons produced from the beams of 10^{12} or 10^{13} protons incident upon fixed metal targets.

A beam of 10^6 antiprotons colliding with a much more tenuous target, presented by a counter-rotating bunch of protons, would not be sufficient to produce observable rates of W and Z's. It takes at least a day, at the normal 2.4 s repetition rate of the PS, to accelerate enough pulses of 26 GeV protons to make enough antiprotons. Antiprotons have to be stored in a small storage ring—a sort of magnetic bottle—for many hours. To collect as many antiprotons as possible, the collector ring needs a large aperture, but its large aperture overflows in a few pulses. To stop this 'bottle' overflowing, its contents have to be continuously concentrated—rather as steam is condensed into a smaller volume. The condensing process is called 'cooling'. Expressed in the six dimensions of phase space, the factor of concentration has to be greater than 10^9. In terms of signal amplification, this is a task rather like detecting a match struck upon the dark side of the moon. Each of the 20 000 antiproton bursts is collected in a small storage ring, the Antiproton Accumulator (AA). CERN's AA ring was designed for 3.5 GeV—the best energy to collect the antiprotons produced by 26 GeV protons from CERN's PS (Billinge 1984).

In the early 1980s a new technique, called stochastic cooling, became available to make this possible (van der Meer 1972). Stochastic means 'to aim for a target and sometimes miss' and this is the essence of the method. One must arrange for the number of 'misses' to be statistically smaller than the number of 'hits'. Of course, the 'hits' are not physical collisions but manipulations in which the amplitude of a particle's motion is reduced. We shall explain the principles of this process, but the reader is recommended to study van der Meer's papers for many of the details (van der Meer 1984).

12.2 Stochastic cooling

12.2.1 Transverse cooling of a single particle

Figure 12.1 shows a particle as it performs betatron oscillations about its closed orbit. By now the reader is familiar with the elliptical trajectory in phase space whose area is called its 'emittance'. It passes between the parallel plates of a fast beam position monitor whose output, proportional to its displacement at the instant of passage, is amplified and sent across the ring to two deflecting plates. These apply an angular divergence increment to realign the particle as it crosses the closed orbit. This reduces its oscillation amplitude and the emittance of the beam to zero. The particle is travelling almost with the velocity of light but, even allowing for the speed of signal transmission in cables and delay in the amplifier, the distance along the chord can be much shorter than the path which the particle follows, and the signal and the particle can be made to arrive together.

The betatron phase advance between the pickup and deflector should be close to an odd multiple of 90°, so that the displacement is compensated by a divergence increment. If the particle arrives at a phase in its motion which is not a maximum, there will be a residual component after the first turn, but this can be corrected on subsequent turns.

Now, of course, the particle will be accompanied by many neighbours and it is not immediately obvious that their signal will not swamp that due to the particle's displacement.

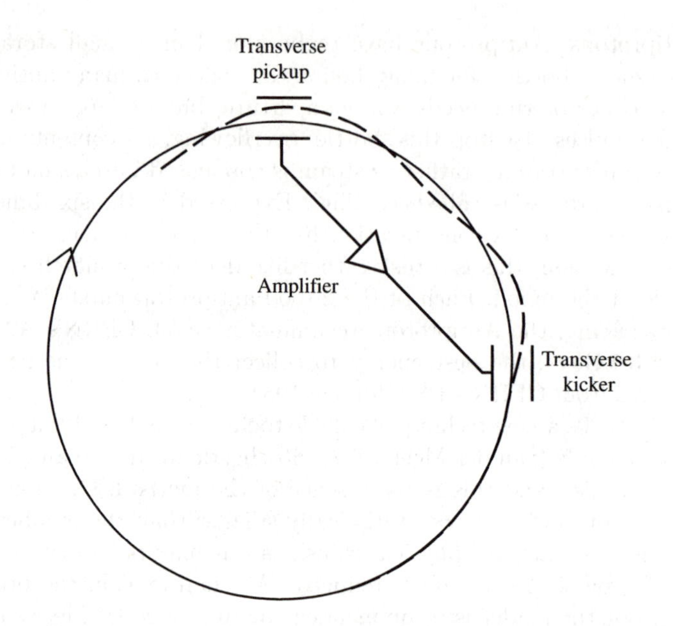

Fig. 12.1 Position pickup signal is amplified and used to deflect the particle.

12.2.2 Cooling a large number of particles

In Fig.12.2(a) we see a histogram of a sample of ten particles which pass through the cooling pickup together. The horizontal coordinate is the displacement of the particles at a pickup. One among the particles is a rogue and displaced far to the right. The pickup produces a signal which is proportional to the mean displacement—the centre of charge—which is inevitably biased in the direction of the rogue.

On the next turn (Fig. 12.2(b)) the rogue has moved somewhat towards the axis. However, the centre of charge has now been restored to the axis by the kick and the resultant pickup's signal is zero. The process develops no further, as long as the same ten particles pass through the pickup together. Then the cooling stops.

However, there are many other particles in the circulating beam and by virtue of their different momenta and revolution frequencies, they can overtake the rogue and leave or join the sample. Companions of the rogue on the first turn may drift out of the sample seen by the pickup. This 'mixing' is essential if the cooling is to continue. In Fig. 12.2(c), the rogue has another set of companions whose centre of charge may be in its direction or in the opposite direction. At first sight it might be thought that any further corrections applied would average to zero without any cooling effect on the rogue but in Fig. 12.2(d) we see how after a number of turns the rogue is progressively brought towards the centre of distribution.

The rate of mixing of a rogue with its companions is determined by the slip factor, $\eta = (1/\gamma^2 - 1/\gamma_{\mathrm{Tr}}^2)$, and the ring's lattice must be designed to make this

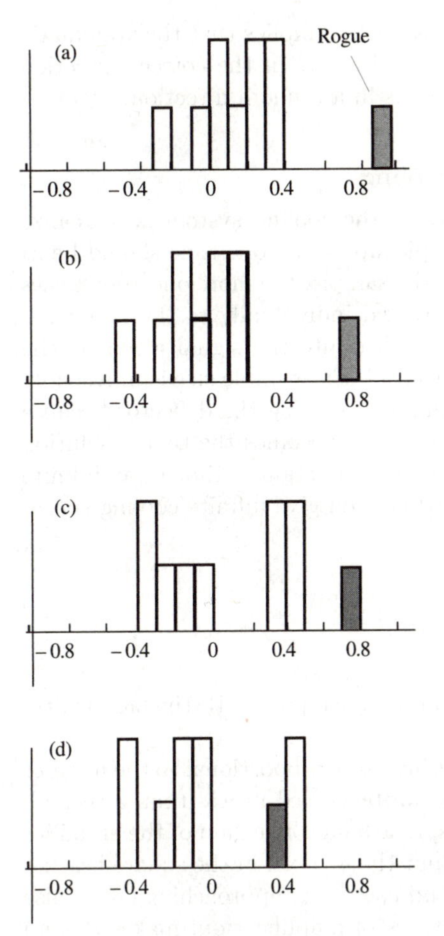

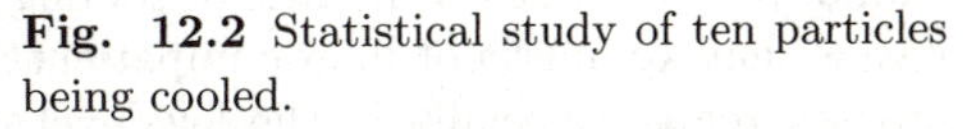
Fig. 12.2 Statistical study of ten particles being cooled.

large but not so large as to destroy coherence between the pickup and corrector. A light-hearted analogy may help persuade the reader that this is not a happy accident in the simulation. Suppose a brigade of soldiers, whose waist measurements have a broad distribution, receive a set of uniforms which are all of medium size. The commandant hits upon an ingenious solution. At the entrance to the canteen he places a corporal who judges the weight of each soldier and sends a signal to the serving counter to give that man double or half portions. After some adjustment of the signalling delay, the message is synchronized with the non-conforming soldier. However, inevitably his companions in the queue receive the same, but often inappropriate, treatment. But so great is the scramble to 'fall out' for meals that he is never with the same group of people. His companions of the day can expect tomorrow to be with others who may be either fat or thin. On an average their portions will be normal, but the fat man never escapes the instructions from the orderly and every time gets less to eat. The thin soldiers are also systematically fattened until one day they all fit the new clothes. The same argument, applied to the ensemble of particles, implies that the systematic correction applied to the rogue particle is consistently in the correct direction while the effect on the well-behaved particles is in a random direction.

12.2.3 Gain, noise, and the rate of cooling

The key to understanding the performance of the cooling system is to realize that the length of the 'sample' seen by the pickup at any one time should be as short as possible. The fewer the particles in the sample, the more one approaches the ideal case in which each particle is corrected individually at the first pass. A large number of particles in the sample will dilute the signal given by the rogue particle and increase the time taken to cool. The sample size is universally proportional to the bandwidth W of the amplifier driving the deflector because the bandwidth, transformed into the time domain, becomes the time resolution of the pickup. A resolution corresponding to delta function in time or an infinite bandwith would detect every individual particle and give infinite cooling speed. The expression for the rate of cooling is

$$\frac{1}{\tau} = \frac{2W}{N}[2g - g^2(1+\rho)] \approx \frac{2W}{N},$$

where g is the gain, $g_{\text{opt}} = 1/(1+\rho)$, ρ the noise/signal power, W the bandwidth, and N the number of particles.

This indicates that the cooling rate is just inversely proportional to the number of circulating particles. A million particles can be cooled in less than a second; 10^{12} particles, sufficient for useful collisions, take a day. The gain of the amplifier will, of course, increase the rate of cooling but the g^2 term in the square bracket means that we must not be too impatient and use a gain approaching one. Noise in the circuit, especially in the low-level stages of amplification, makes this an even more severe limitation. Moreover, there is a practical limit to the product of gain and bandwidth for amplifiers which constrains the cooling rate.

In general, the way in which the gain, bandwidth, and noise scale argues for a frequency as high as possible yet consistent with the aperture of pickups and kickers whose dimensions must be big enough for the beam to pass through.

12.2.4 The position detectors

Typical high-frequency pickups for cooling are slotted transmission lines working in the GHz region (Fig. 12.3). The one shown has four rows of slots on each side of a rectangle. Difference signals can be used to pick up betatron oscillations and sum signals to detect momentum or arrival time changes. Kickers have a configuration similar to the pickups. The correction of betatron motion is fed in push–pull mode and that of momentum deviation is fed in common mode. Larger aperture devices can be made with ferrite-loaded transmission line kickers, but these are limited in frequency and used only for cooling weaker beams.

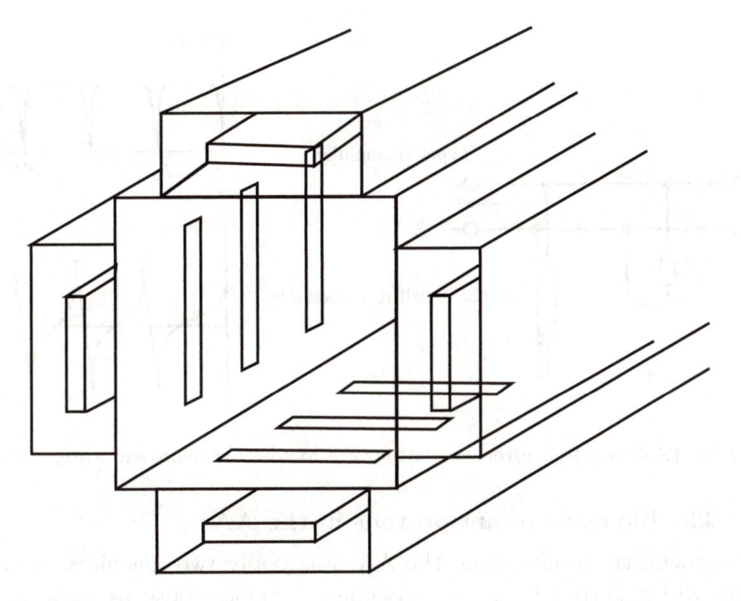

Fig. 12.3 A high-frequency pickup for stochastic cooling.

12.2.5 Momentum cooling

We can measure the momentum of a particle by detecting its revolution frequency and correct it by applying an accelerating voltage to a cavity. Naturally, we need more than one turn to measure a revolution frequency, so the picture is less obvious than the transverse case. It is probably easier to imagine what is happening in the frequency domain. The pickup is continuously responding to a spectrum of frequencies. Let us say for the moment that these are the revolution frequencies of individual particles of different momenta. We must filter its output so that it is blind to particles with the correct revolution frequency. Any particle with a lower momentum gives a higher-frequency signal (above transition), which escapes the filtering and, when applied to the accelerating cavity, will accelerate and correct the particle's momentum until it falls back into the desired range defined by the filter. The filter must be designed so that signals above its passband are 180° out of phase with those below. In fact, a simple resonant circuit has just this characteristic. This ensures that low-frequency signals provoke an accelerating voltage while high-frequency signals decelerate. Again the statistics of the parade ground apply and there must be the right amount of mixing.

Of course, not only the fundamental revolution frequency is generated but also a comb of higher harmonics, each of which is useful. Filters which deal with a comb of frequencies can be constructed from resonant delay lines whose electrical length corresponds to half a wavelength at the fundamental frequency (Fig. 12.4).

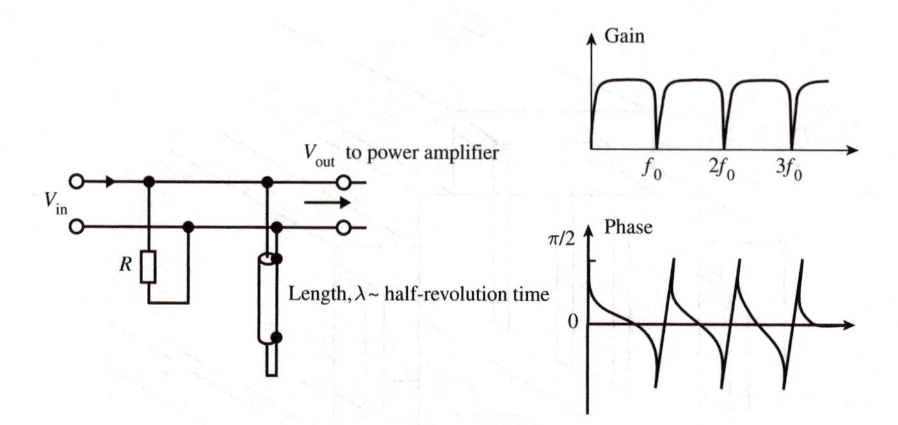

Fig. 12.4 A notch filter consisting of a low-loss transmission-line stub.

12.2.6 The life cycle of antiprotons in the AA

CERN's stochastic cooling ring, the AA, was really two machines in one—two machines which Fermilab, in its subsequent scheme, chose to make physically distinct. The injected beam, with as large an emittance and momentum spread as can be accepted, rides in the outer half of the vacuum chamber. On the inside of the vacuum chamber, circulating at a different momentum and radial position, is the cooled stack of antiprotons which has been accumulated over many hours of regular pulsing. The life cycle of antiprotons is shown in Fig. 12.5.

Liouville's theorem prevents us from superimposing the injected particles directly on the stack. In practice, an attempt to do so with the kicker which inflects the new pulse onto its orbit would eject the stack.

There are three steps in bridging the gap between injection orbit and stack. They are pre-cooling, r.f. deceleration, and stack-tail cooling. The pre-cooling is designed to reduce drastically the momentum spread of the injected pulse by an order of magnitude so that a rather weak r.f. system can capture, decelerate and deposit it gently in the tail of the stack in time for a new pulse to take its place after 2.4 s.

Once in the tail, the 'stack-tail' cooling system takes over, detecting that the particles have a lower frequency than the stack because their momentum and mean radius are different, it applies a signal to a cooling kicker which decelerates them towards the stack frequency. On their way their horizontal and vertical betatron amplitudes are reduced by stack-tail transverse systems.

Once in the stack a third set of cooling systems take over: the high-frequency core cooling systems designed for intense beams.

12.3 Electron cooling

Electron cooling was first developed at Novosibirsk under the aegis of Budker (1967). A beam of heavy particles, antiprotons or ions, for example, travels

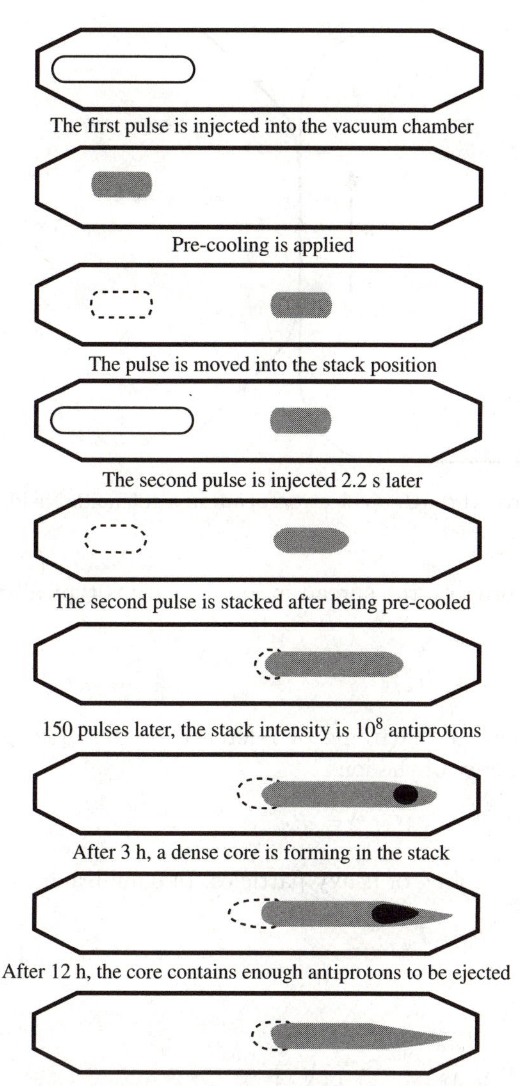

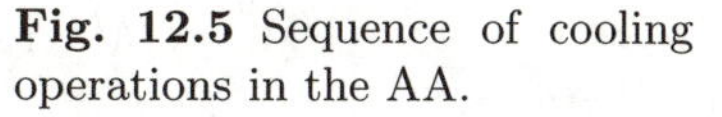
Fig. 12.5 Sequence of cooling operations in the AA.

through a beam of electrons with the same velocity transferring its energy to the electrons (Fig. 12.6). Seen in the co-moving system of the two species of particles, this is simply an equipartition of energy as the two species experience their mutual electromagnetic fields. It is a much faster process than stochastic cooling but most effective when the proton or antiproton beam is circulating at low energy.

Although the principle is akin to the equipartition of energy, the electron gas is constantly renewed, and the ion temperature will converge on the electron beam's temperature.

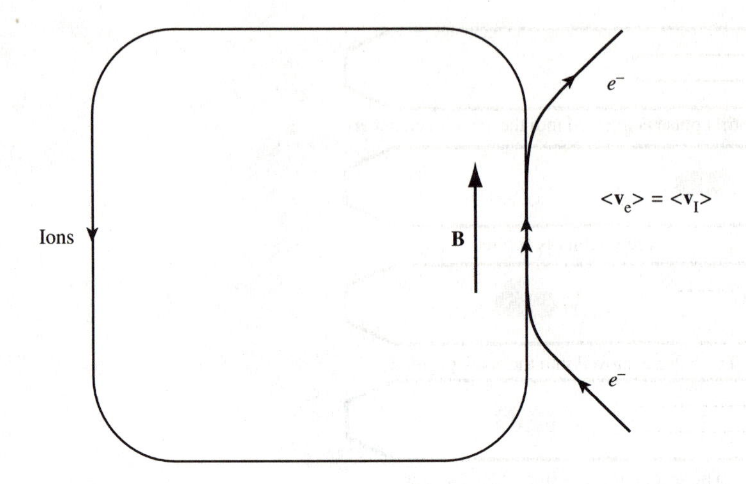

Fig. 12.6 Ions are cooled as they travel through an electron beam in a solenoidal field.

In the beginning of the cooling process, the spread in ion velocity is typically larger than the electron velocity:

$$\langle v_{\mathrm{I}}^2 \rangle \geq \langle v_{\mathrm{e}}^2 \rangle,$$

where the brackets signify averages in the frame of reference of the moving particles. So initially the kinetic energy of the ion

$$T_{\mathrm{I}} \equiv \tfrac{1}{2} M \langle v_{\mathrm{I}}^2 \rangle > \tfrac{1}{2} M \langle v_{\mathrm{e}}^2 \rangle \equiv T_{\mathrm{e}},$$

since we only consider the electron cooling of heavy particles. In equilibrium,

$$T_{\mathrm{I}} = T_{\mathrm{e}}$$

and

$$v_{\mathrm{I}}^{\mathrm{rms}} = \sqrt{\langle v_{\mathrm{I}}^2 \rangle} = \sqrt{\frac{m}{M}} v_{\mathrm{e}}^{\mathrm{rms}} \approx \frac{1}{43} \sqrt{\frac{1}{A}} v_{\mathrm{e}}^{\mathrm{rms}};$$

so, finally, the velocity spread of the ions will be much smaller than the velocity spread of the electrons.

The beam of electrons must be accelerated and focused into a precisely parallel beam and decelerated in a collector after interacting with the ions (Fig. 12.7).

Readers who wish to know more about electron cooling are recommended to follow Møller (1992), where one may find an expression for the cooling time. Electron cooling is most effective at low energy and is used in a number of small storage rings for ions and antiprotons: LEAR and AD at CERN. The need to decelarate antiprotons before cooling ruled out its use in the CERN and Fermilab antiproton sources.

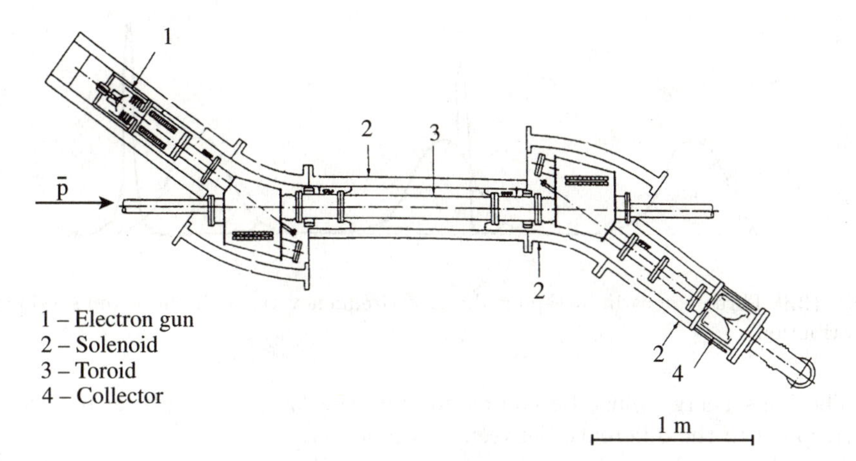

Fig. 12.7 Electron cooling apparatus.

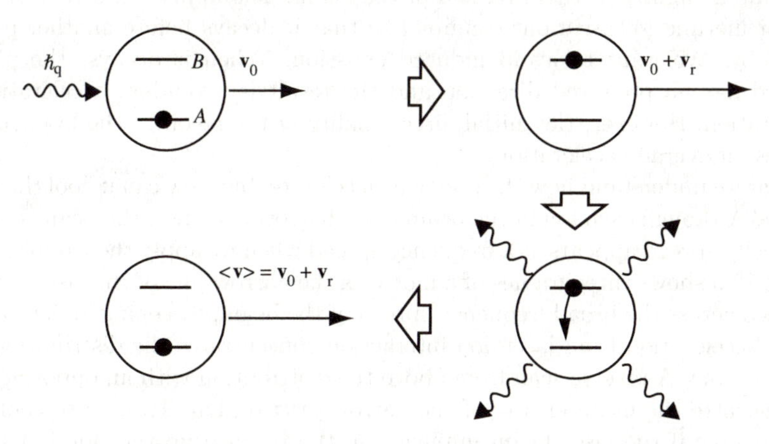

Fig. 12.8 Directional absorption and isotropic emission.

12.4 Laser cooling

This is a fascinating practical application of modern physics (Møller 1992). The idea can be applied to a beam of ions circulating in a storage ring. The ions must be of a particular species and charged state chosen to have a ground state A and upper energy level B, separated by an energy corresponding to a tuneable laser (Fig. 12.8.) An example is a 100 keV $^7Li^+$ beam whose transition from 3S(1s2s) to 3P(1s2s) can be excited by a laser of 5485 Å. This is within the tuning range of a continuous wave dye laser.

If the laser and the beam travel together in a straight section of a storage ring, the Doppler-shifted frequency experienced by the ion is

$$\omega' = \gamma\omega(1 - \beta\cos\theta),$$

where β and γ are Lorentz variables and θ is the angle between the beams.

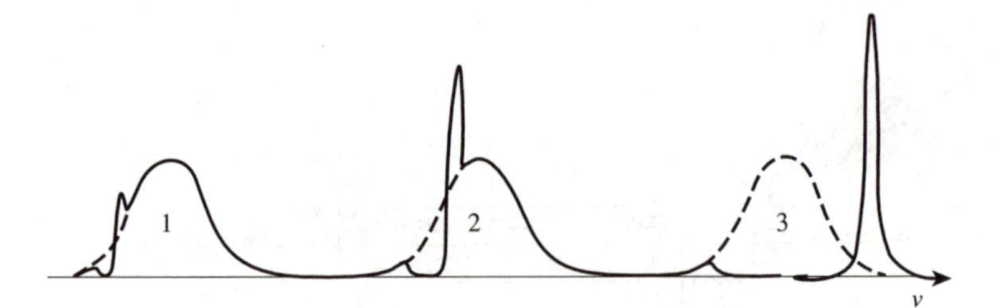

Fig. 12.9 Three stages in sweeping the laser frequency through the beam's velocity distribution.

The ion's energy must be chosen to shift the laser frequency ω to exactly correspond to the difference between energy levels.

If the laser and the ion are in resonance, a photon is absorbed and the ion accelerated slightly in the direction of the beam. The upper state B must have a short lifetime (42 ns in our example) so that it decays before another photon passes by. We want to avoid induced emission. When it decays, the photon emitted has no preferred direction and the result is a random acceleration or deceleration. However, the initial directionality of the recoil to the laser photon ensures an overall acceleration.

So far we understand how the laser can accelerate, but how can it cool the beam of ions? Although each ion in the beam has a narrow response, the beam's energy or velocity spread appears as a frequency spread when we apply the Doppler shift. Figure 12.9 shows three frames of a movie as the narrow line of the laser is tuned to sweep across the broad frequency spread of the beam. Sweeping all ions before it, the acceleration from laser–ion interaction concentrates the distribution into a narrow line. A reverse scan from above the distribution with an opposing laser can re-centre the mean energy of the narrow distribution. Transverse cooling is also possible if the laser beam impinges on the beam from the side but this is inefficient.

Laser cooling can be extremely fast though the limited number of suitable combinations of ions and lasers restricts its application to certain energies and species of particles.

12.5 Ionization cooling

Muon colliders today offer a route to extend circular lepton accelerating storage rings to energies of several TeV without the annoying energy loss from synchrotron radiation that presently limits LEP.

Muons, and their parent pions, must be collected and cooled in the fashion of antiprotons but in a time which competes with their rapid decay. To reach reasonable intensities, the cooling time has to be a few milliseconds. Single pass cooling is the only solution. Particles passing through an energy absorbing plate lose momentum in the direction of their trajectory in exactly the same way as a

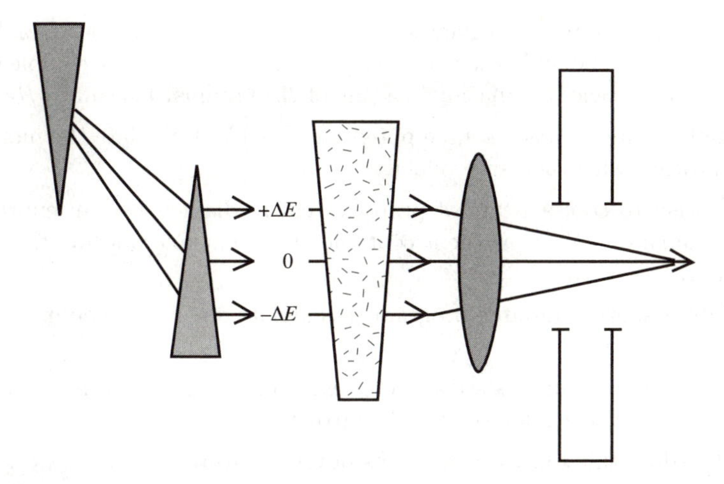

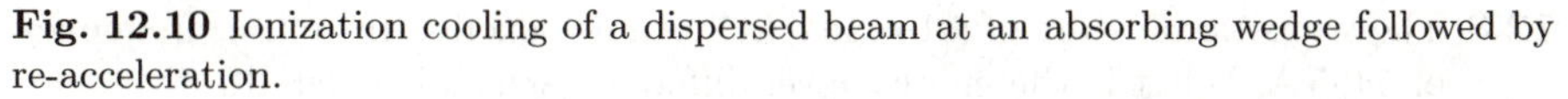

Fig. 12.10 Ionization cooling of a dispersed beam at an absorbing wedge followed by re-acceleration.

relativistic particle emitting a quantum of synchrotron light. An r.f. cavity following the absorber replaces the longitudinal momentum but not the transverse. This is just as portrayed in Figs 8.4 and 8.5, when we explained transverse damping for electron synchrotrons. The result is a reduction of transverse emittance. Such ionization cooling would not work for electrons because of bremsstrahlung and hadrons because of strong interaction with the absorber, but it works for muons.

Longitudinal cooling may also be achieved with an absorber—the process is called emittance exchange (see Fig. 12.10). If a beam's energy or momentum distribution is spread by dispersion across a wedge of solid material, the high-energy particles lose more energy from the wedge than their low-energy companions, thus 'cooling' the energy spread. Unfortunately, the result is simply that the energy spread is reduced at the expense of an inflation of transverse phase space, but transverse cooling can be applied again to remove this.

In plans for a muon collider, now being strenuously studied in the US, many alternating stages of wedge absorbers, ionization cooling and emittance exchange are envisaged. Initial simulation studies give encouraging results.

Exercises

12.1 Use the following expression for the cooling rate:

$$\frac{1}{\tau} = \frac{2W}{N}[2g - g^2(1 + \rho)],$$

and obtain an optimum value for the gain g in the presence of a signal-to-noise ratio ρ. Write down the expression for the cooling rate for this optimum gain.

12.2 A cooling system is designed with a central frequency of 300 MHz. What is the sample time and how many particles will be in any one sample of 10^6 a beam of particles circulating in a ring of 25 m radius? (Assume $v/c = 0.96$.)

12.3 An electron gun has a source potential of 60 kV. Calculate the momentum of protons with the same velocity.

12.4 We wish to cool a beam of protons 5 cm in diameter to an emittance of 40π mm mrad. What is the acceptable alignment tolerance on the electron beam?

12.5 Write a short computer program to simulate stochastic cooling with and without mixing.

12.6 If the proton beam has a transverse emittance of 2π mm mrad, what would be its transverse velocity at a β of 10 m?

12.7 What does this represent in terms of temperature?

12.8 The transitional state of a 100 keV Li^+ beam is excited by a laser frequency of 5485 Å. What is the energy level difference which is excited?

13
Applications of accelerators

13.1 Introduction

The reader will perhaps be surprised to hear that there are about 10 000 accelerators in the world and the vast majority are not built to study the fundamental particles of matter but are put to more 'practical' purposes.

Table 13.1 is from a world-wide study by Scharf and Chomicki (1996) who list a total of 112 accelerators of more than 1 GeV. Only one-third of these are dedicated to high energy physics while the rest are mainly synchrotron light sources. There are a further 5000 accelerators of lower energy for medical purposes: radiotherapy, biomedical research, and isotope production. A comparable number is deployed in industry, mainly as ion implanters and for surface treatment. In fact, more than 99% of the world's accelerators have been built for use outside the discipline of particle physics.

13.2 Industrial processes using accelerators

Electron beams create electron showers which degrade to lower energy, where they excite chemically active sites. These can break up biological molecules in an organism, rendering it innocuous, or promote new bonds which polymerize

Table 13.1 Particle accelerators—world-wide (after Dearnaley 1987)

Category of accelerators	Number in use
(1) High-energy accelerators of more than 1 GeV	112
Biomedical accelerators	
(2) Radiotherapy	>4000
(3) Research including biomedical research	800
(4) Medical radioisotope production	~200
(5) Accelerators in industry	~1500
(6) Ion implanters	>2000
(7) Surface modification centres and research	~1000
(8) Synchrotron radiation sources	~50
Total in 1994	10 000

Table 13.2 Electron irradiation (after Dearnaley 1987)

Industries	Processes	Products
Chemical	Cross-linking	Polyethylene
Petrochemical	Depolymerization	Polypropylene
	Grafting	Co-polymers
	Polymerization	Lubricants
Electrical	Cross-linking	Building
	Heat shrink memory	Instruments
	Semiconductor modification	Telephone wires, power cables, insulating tapes, shielded cable splices, Zener diodes, ICs, SCRs
Coatings adhesives	Curing	Adhesive tapes
	Grafting	Coated paper products
	Polymerization	Wood/plastic composites, veneered panels, thermal barriers
Plastics	Cross-linking	Food shrink wrap
Polymers	Foaming	Plastic tubing and pipes
	Heat shrink memory	Moulded packaging forms
Rubber	Vulcanization	Tyre components
	Green strength	Battery separators
	Graded cure	Roofing membrane

and harden plastics. A large number of industrial processes make use of electron beams; many everyday objects rely on electron beam hardening—among them are computer disks, shrink packaging, motor car tyres, cables, and plastic hot-water pipes. Table 13.2 lists these processes and the end products. The energy and intensity of the beams required vary greatly. The energy is usually determined by the depth of penetration required and ranges from surface treatment, needing only a few hundred keV, to treatment of bulk material, where a beam of several MeV is needed to penetrate tens of centimetres of material.

13.2.1 Sterilization

Particle beams may be used for applications that include disinfecting drinking water, treatment of solid wastes, removal of noxious substances, treatment of waste gases, medical sterilization, and preservation of food.

Sterilization of foodstuffs is still one of those issues which the general public finds difficult to accept, yet the potential benefits are impressive. Doses of a few hundred Gy will render most insect pests sterile and lead to their death within a few days (Fig. 13.1), preventing the deterioration of bulk grain, vegetables, and fruits. A dose of 200 Gy will arrest their germination. Cooked food can be stored almost indefinitely at room temperature if it is packed and irradiated with a few hundred Gy and less benign artificial preservatives are no longer needed.

Fig. 13.1 Weevil in a sack of grain.

There are also some sterilization processes where accelerators can be used without exciting public debate. The accelerator is an alternative to the autoclave to sterilize surgical instruments and laundry. Also the preservation of foodstuffs for animals and the production of fertilizers for crops are generally perceived to be sufficiently remote from human ingestion, at least as far as the effects of radiation are concerned. Another unquestionably benign use of radiation is in destroying the bacteria infesting the detritus of the operating theatre. Even the sludge of sewage could be usefully incorporated in some products after radiation to kill pathogenic micro-organisms, though recently public opinion has turned against this. The required dose is of the order of 10 kGy.

13.2.2 Doses

The doses required for the various applications span a large range. In Table 13.3 we see that the current and power requirements for medical purposes are very low but at the other end of the spectrum, disinfecting sewage and drinking water on a large scale require quite powerful installations. A 12 MeV accelerator to treat the drinking water for a town of 100 000 people would have to deliver a beam power of 600 kW.

13.2.3 Ion implantation in semiconductor manufacture

Most of the applications mentioned above use electrons, but simple DC accelerators are used in great numbers in industry to accelerate ion beams of low

Table 13.3 Dose requirements for various radiation effects (after M. R. Cleland)

Radiation effect	Dose requirements
Radiography (film)	1.0–10.0 mGy (0.1–1.0 rad)
Human lethal dose (LD_{50})	0.4–0.5 Gy (400–500 rad)
Sprout inhibition (potatoes, onions)	100–200 Gy (10–20 krad)
Potable water cleanup	250–500 Gy (25–50 krad)
Insect control (grains, fruits)	250–500 Gy (25–50 krad)
Waste water disinfecting	0.5–1 kGy (50–100 krad)
Fungi and mould control	1–3 kGy (100–300 krad)
Food spoilage bacteria	1–3 kGy (100–300 krad)
Municipal sludge disinfecting	3–10 kGy (300–100 krad)
Bacterial spore sterilization	10–30 kGy (1–3 Mrad)
Virus particle sterilization	1–30 kGy (1–3 Mrad)
Smoke scrubbing (SO_2 and NO_x)	10–30 kGy (1–3 Mrad)
Ageing of rayon pulp	10–30 kGy (1–3 Mrad)
Polymerization of monomers	10–50 kGy (1–5 Mrad)
Modification of polymers	50–250 kGy (5–25 Mrad)
Degradation of cellulose materials	100–500 kGy (10–50 Mrad)
Degradation of scrap Teflon®	0.5–1.5 MGy (50–150 Mrad)

energy and are an essential tool in the manufacture of semiconductors. A typical semiconductor production process might contain 140 operations, of which 70 involve the implantation of ions in the crystal lattice of the semiconductor. The implantation of ions at specific lattice sites and the creation of defects is a highly developed technology. The depth of the implant is controlled by choosing the ion energy which is usually between 2 and 600 keV. The species of the ion is selected by standard mass analysis techniques. Similar methods are now applied for the manufacture of superconducting materials where implantation is used to 'pin' atomic planes.

13.2.4 Surface hardening with ions

Ions are also used for the surface treatment of metals in the engineering industry. Tungsten, chromium, titanium, tantalum, nitrogen, boron, and other ions may be implanted to harden the surface of steel components such as ball bearings and cutting tools (Fig. 13.2), and to avoid corrosion (Grob *et al.* 1996).

Unlike more conventional surface-hardening techniques with chemicals and high-temperature furnaces, ion beams do not heat the surface and further annealing is not required. The ions can be of metallurgically 'forbidden' atoms and can be implanted in a surface layer which will avoid subsequent fissures. Typical applications are in the manufacture of artificial hip and knee joints (Fig. 13.3) and in the manufacture of control and fuel rods for nuclear reactors. The density of implantation for hardening purposes is about 100 times that used in the manufacture of semiconductors and the accelerator must produce currents in the range of 5–10 mA at energies in the range 50–200 keV. The flux required for ion

Fig. 13.2 Hardening gears with ions (artist's impression).

implantation is considerable and the number of ions which are deposited is in the range 10^{15}–10^{17}. The latter number represents as much as 10% of a layer of material several thousand angstroms in thickness.

13.2.5 Precision machining and membrane manufacture

Ion beams may be used as precise tools for machining plastic surfaces to a depth which far exceeds the transverse dimensions of the surface features. In Fig. 13.4 we see how ion track etching is used to produce an extremely fine filter from a polymer foil. It is possible to make membranes with track diameters from 10 μm down to 10 nm and densities from 1 to 10^9 pores per cm^2.

13.3 Types of accelerator used in industry

13.3.1 Electrostatic single-stage accelerators

We start at the low-energy end of the spectrum. Many of the accelerators used in industry for surface treatment require only a low energy—often less than 750 keV.

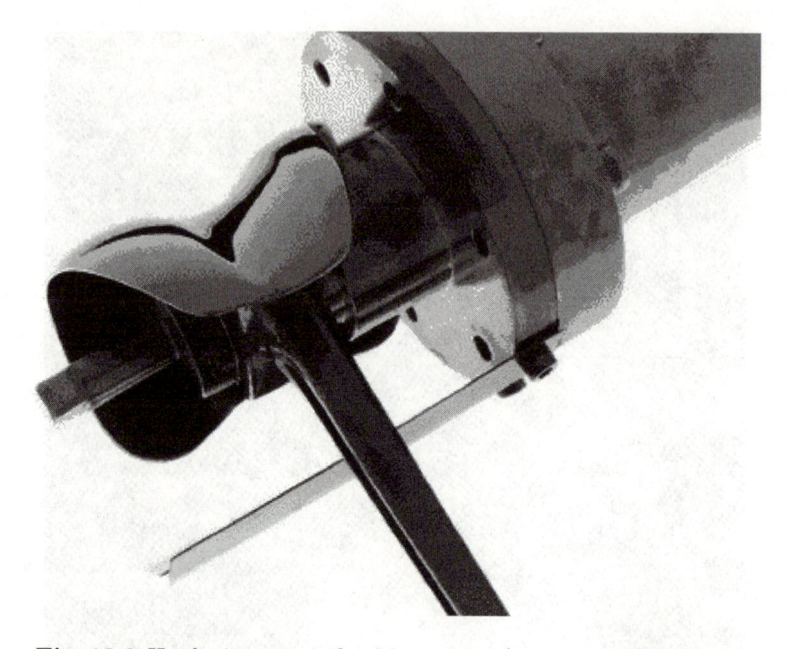

Fig. 13.3 Hardening an artificial knee joint (courtesy of GSI photo).

Fig. 13.4 Filter micro-machined with ions (courtesy of GSI photo).

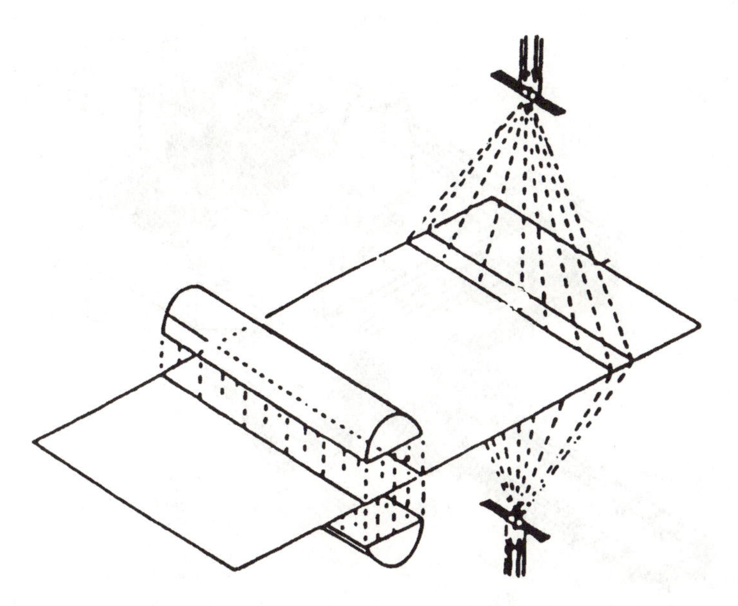

Fig. 13.5 Methods of surface treatment (after Scharf 1996).

Fig. 13.6 Sterilization of surgical supplies (CERN courier photo).

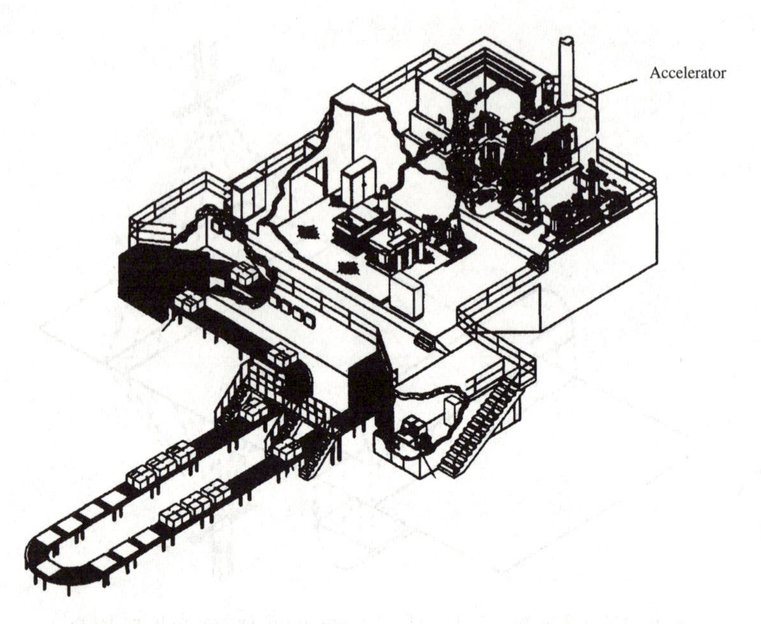

Fig. 13.7 An industrial sterilization facility (CERN courier photo).

This can be achieved by a simple electrostatic accelerator. A DC potential is applied to a wire anode and electrons extracted through a slot in a co-axially mounted cylindrical cathode. Such a strip treatment setup is shown in Fig. 13.5. An alternative technique is where the beam emerging from a simple electrostatic accelerator is swept over the surface of a moving belt.

Where low intensity is needed and when the energy must be high enough to penetrate many grams of material, electron linear accelerators of a few MeV are often the preferred solution. We need an electron beam energy of 5–10 MeV to penetrate a significant thickness of material, say 20 cm of plastic. In Fig. 13.6 we see boxes of medical and surgical produce passing on a conveyor below the accelerator. The beam, descending vertically, is swept from side to side with an oscillating dipole to cover the target volume in the manner of Fig. 13.5. All of this must take place in a well-shielded vault separated from the outside world by chicanes (Fig. 13.7) (Drewell *et al.* 1996).

13.3.2 Ion accelerators

Ion sources usually involve the bombardment of a gas or vapour with electrons. Often the electrons come from an arc discharge between two electrodes and the ions emerge from a slit in a cathode aligned with the arc. Once through the slit, the beam is accelerated towards a larger hole in an extraction electrode. This beam must be neutralized with an 'electron gas' to minimize defocusing due to space-charge forces. Usually, the ion source must be at the positive high-voltage terminal of the accelerator so that the accelerated beam emerges at ground

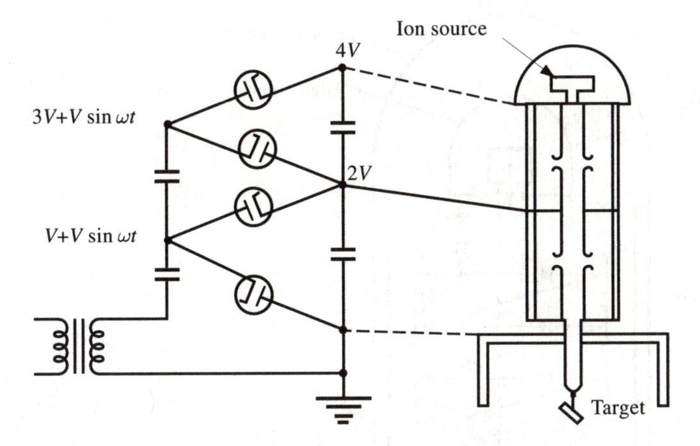

Fig. 13.8 The principle of a Cockcroft–Walton machine (after Scharf 1996).

potential. This may be followed by r.f. quadrupoles and a linear accelerator structure specially designed for low-velocity particles.

13.3.3 Cockcroft–Walton and Dynamitron®

Linacs, although favoured for electrons in the 1–5 MeV range, become complicated if they are to accelerate particles of different charge-to-mass ratios at velocities below that of light. Protons or ions may be accelerated with Cockcroft–Walton or Dynamitron® accelerators to higher energies than are possible with a simple electrostatic gun. These machines can also be used to accelerate electrons. Figure 13.8 shows the principle of the Cockcroft–Walton circuit: a chain of diodes which alternate in polarity and are capacitively coupled to each other. A relatively modest alternating voltage, V, applied across the lower diode is multiplied by the number of the diodes in the chain. Each cell acts as a full-wave rectifier to the alternating voltage, and their DC voltages add up in series across an accelerating column.

The dynamitron (Fig. 13.9) uses a similar diode column but the alternating, r.f. potential is applied in parallel to each diode from a cylindrical capacitively coupled r.f. electrode. Such DC accelerators are filled with insulating SF_6 gas under pressure and large-dimension vessels separate the anode and cathode.

13.3.4 van de Graaff accelerators

Although Cockcroff–Walton and Dynamitron® accelerators are used extensively up to a few MeV, a different kind of DC machine, invented by van de Graaff in the early 1930s, takes over above this energy and may be used up to about 15 MeV. In a van de Graaff accelerator, a moving belt transports charge to a high-voltage terminal, which forms one end of an accelerating column. In a tandem configuration, negative ions produced by adding loosely bound electrons

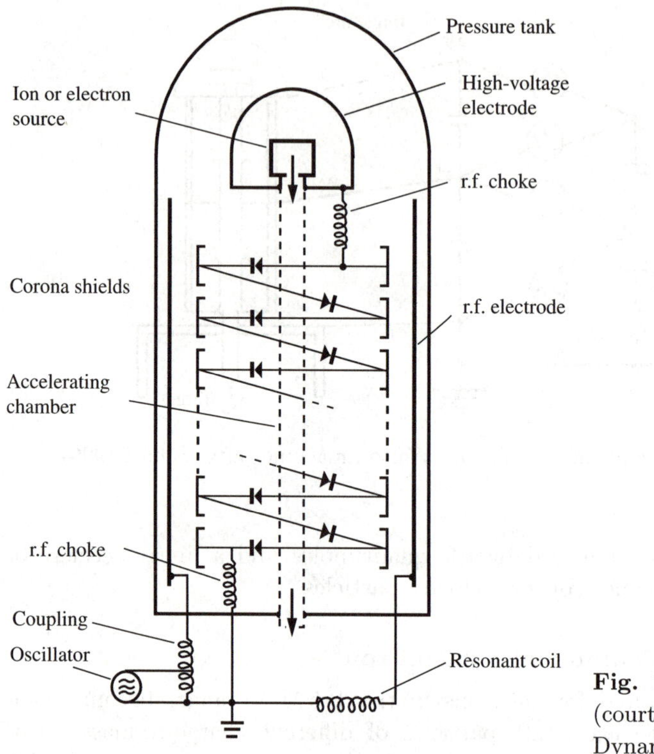

Fig. 13.9 The dynamitron (courtesy of Radiation Dynamics Inc).

to the neutral atom, may be accelerated from ground to a positive terminal, stripped, and further accelerated as positively charged ions towards a ground potential. This principle doubles the energy of the emerging beam for any given terminal voltage. The pressure vessel is often filled with SF and the current is typically in the range 0–100 μA. In Fig. 13.10 we see a horizontally mounted van de Graaff machine from which two spectrometer arms emerge. This particular setup is used to activate trace elements in a sample and identify the elements from their reaction products analysed in a mass spectrometer.

13.4 Medical applications

13.4.1 Isotope production

Accelerators, mainly cyclotrons, produce about 20% of the radio-pharmaceutical materials injected into patients and about 35 of the 200 of the world-wide inventory of cyclotrons are used for this purpose. Typically, these cyclotrons accelerate protons in an energy range up to 40 MeV in beams of 50–400 μA. They must be designed to be compact, reliable, and to produce high-intensity extracted beams with a minimum of human intervention (Bechtold 1996; Lewis 1996). Remote handling of the targets is essential at these intensities and must ensure a

Fig. 13.10 Horizontally mounted van de Graaff (CERN courier photo).

speedy transfer into the often automated procedure for radiochemical extraction, dispensing, labelling, packaging, and delivery of short-lived products.

The isotopes commonly have a half-life of about three days. When injected, their activity should be low enough to keep the effective dose below 5 mSv (0–5 rad) yet provide diagnostic information with optimal γ-ray imaging in a 15–30 min session.

13.4.2 Positron emission tomography

This diagnostic technique uses short-lived isotopes in conjunction with imaging systems based on particle detection techniques. The isotopes emit positrons, which are detected as back-to-back γ's and which may be projected back to locate the emission point. The isotopes can be incorporated into biochemical molecules which find their way to particular sites in the body, revealing a three-dimensional picture of biochemical activity. So short is the half-life of these positron emitters (~20 min) that the small cyclotron which produces them and the automated synthesis system have to be installed in the room next to the patient.

13.4.3 Therapy

In the past, radiotherapy made extensive use of radium needles or 2 MeV γ-rays from cobalt 'bombs'. Electron linear accelerators in the range of 15–20 MeV are currently used to produce X-rays to reach deep tumours (Gahbauer and Wambersie 1996). The beams must converge on the tumour from several directions if healthy tissue is to be spared. The margin between a dose sufficient to destroy malignant cells and that low enough to allow healthy tissue to regenerate is only 10% or 20%.

13.4.4 Proton therapy

Attempts to spare intervening tissues are not helped by the exponential decay of X-ray intensity as it penetrates the body (Wilson 1946). Protons offer a better solution since they deposit most of their energy in a sharp 'Bragg' peak, leaving intervening tissue relatively unharmed and completely sparing sensitive organs just beyond the tumour site (Sisteron 1996).

The ideal energy for protons is 200 MeV, sufficient to reach any internal organ. Proton synchrotrons have been constructed for this purpose notably at Loma Linda, PSI, San Diego, USA, and HIMAC in Japan. Others are planned and the alternative of a superconducting cyclotron has led to new projects in Massachusetts, USA, and NAC, South Africa. Cyclotrons with their continuous beam of small emittance enable very precise control of dose and treatment zone.

13.4.5 Ion therapy

Recent research into the density of ionization as ions pass through cells suggests that ions are better suited than protons in 'taking out' large sections of DNA which cannot regenerate. Trials to verify this in a chemical situation are underway at LBL, Berkeley, and HIMAC, Japan. Other synchrotron projects which include the light-ion option are planned at GSI, Darmstadt, and TERA in Italy.

13.4.6 Beam delivery

Millimetre precision is often needed to confine the dose to the tumour. The beam energy may be modulated with a rotating disc of absorber of varying thickness and the beam shape defined by fixed or movable leaf collimators. In the simpler systems, the irradiated zone is made to conform to the tumour by directing the beam from one side, then the other, and also perhaps from above. Complex gantries have been built which direct a horizontal beam from the axis of a rotating wheel structure to the rim and then in, towards the patient at the turn of the structure. These are favoured by proton therapy centres but represent a large fraction of the cost of the facility. Gantries for ion beams are more difficult because of the large emittance and magnetic rigidity.

13.5 Research applications

13.5.1 The high-energy frontier

We are all very familiar with the use of high-energy lepton and hadron colliders for particle physics. The essential questions under study include the search for the Higgs particle thought to be the origin of mass, the stability of protons, and the mass of neutrinos. At the time of writing, accelerator builders are considering using muon storage rings as a precise source of neutrinos as an alternative to solar and atmospheric neutrinos. All this research has an intimate relation with astrophysics and cosmology in re-creating the particles and interactions which prevailed just after the Big Bang and before nuclei were formed from more fundamental particles. The data are essential in simulating cosmological models and to help us understand the mystery of the missing mass in the universe.

13.5.2 Nuclear physics

Equally important, but a little later in the chain of events of creation, are the measurement of interactions of protons, neutrons, and the nuclei as they began to condense out from the primordial soup. Much of this information came, and is still coming, from nuclear physics research with the low-energy van de Graaff accelerators and cyclotrons, which were developed in the early days before and after the Second World War. More recently, heavy ions have been used to probe nuclear structures with the techniques of spectroscopy. Modern research focuses upon nuclei which are unstable and lie far from the mainstream of stable nuclides or which are anomalous in other respects such as superdeformed nuclei with very high-angular momentum.

The availability of ion beams enables us to study nucleus–nucleus collisions, and, with the aid of beams of both high energy and intensity the meson change and quark structure of nucleons—work that is now actively being pursued at CERN's Isolde and at Ganil in France. Returning to even higher energy, there is evidence that CERN's heavy-ion source combined with the SPS collider have been able to establish the conditions in which nucleons dissociate into gluons and quark plasma. Such research will be a principle activity for RHIC at Brookhaven Laboratory and RIKEN in Japan.

13.5.3 Techniques for the analysis of materials with particle beams

The analysis of small samples of material is an essential requirement in many fields of research and can prove useful in fields unrelated to science. Examples range from the dating of archaeological objects such as the Turin shroud, or sequencing Galileo's manuscripts to the detection of explosives and contraband in freight cargoes rolling through the entrance to tunnels. The techniques (Bethge 1993; 1996a,b) are too numerous to describe in detail, but we list some of the more important methods below.

13.5.3.1 Rutherford Backscattering (RBS)

Rutherford backscattering is a technique in which the energy of ions is measured as they are backscattered from a sample by Coulomb interaction. Peaks in the spectrum are a sensitive indicator of particular target nuclei. The cross section is in fact proportional to the Z^2 of the sample. The sensitivity is quite phenomenal and can be as high as 0.1 ppm. A typical probe beam consists of He^+ ions at 2 MeV per charge. The penetration depth, even for these light ions, is only a few microns but heavier ions may be used to improve resolution when thin films are under study. Higher-energy H^{++} ions (4 MeV) can be scattered by the nuclear potential to exhibit resonances which are characteristic of the target nucleus.

13.5.3.2 Particle-Induced X-Ray Emission (PIXE)

In this technique, an ion beam, often protons, is used to excite target atoms so that they emit characteristic X-rays. The method probes a bulk sample but may be made to have high lateral resolution by focusing the incident beam down to a few microns and enabling high-resolution maps of the surface composition to be made and later compared with electron micrographs (Baglin *et al.* 1996). Like RBS, its sensitivity may be as high as 0.1 ppm. To achieve both high resolution and sensitivity, the brightness of the particle beam is crucial. The extreme sensitivity of this technique with its non-destructive character leads to its popularity for detecting anachronistic chemicals in 'ancient' artefacts.

13.5.3.3 Nuclear Reaction Analysis (NRA)

A general class of analytic methods makes use of narrow resonances in the cross section of the reaction of the accelerated particles with different nuclei. A large body of data tabulating these resonances grew up as accelerator beams were used to probe nuclei, and now the more striking features can be used to identify many specific constituent nuclei. Typically, scattering of α particles shows a sudden enhancement at 3.045 MeV when ^{16}O is present in the sample. This is a technique that lends itself to heavy-metallic substrates and is much used in investigating high-T_c superconductors.

13.5.3.4 Elastic Recoil Detection (ERD)

In a technique which complements RBS, one observes the recoil of the heavy target nucleus. It is a well-known feature of the kinematics of such scattering that the recoiling nucleus is scattered with a distribution which becomes more restricted in angle as the mass of the nucleus increases. Alternatively, the time of flight (energy) of the recoil particle can be used to determine its nature.

13.5.3.5 Charged Particle Activation Analysis (CPAA)

This is the most sensitive of the activation analysis methods (1 ppb) for elements like boron, carbon, nitrogen, and oxygen, whose unstable isotopes are positron emitters. Samples are irradiated for times of the order of several half-lives and then, from the decay curves of the various excited nuclei, one may extrapolate back to determine the relative composition at the time of irradiation. The sensitivity of this method lends itself to the study of wear and corrosion. This technique is used to screen luggage and other goods in transit for explosives or drugs.

13.5.3.6 Accelerated Mass Spectroscopy (AMS)

An even more sensitive method (10^{-14}) for relative abundance of isotopes consists in accelerating sample particles in a beam, stripping to remove contamination, and then using momentum analysis to separate the spectrum of masses. Only very small quantities of material need to be sacrificed and the method is particularly appropriate for C^{12}, C^{14} separation in age determination, replacing the earlier β decay counting method (Jianjun *et al.* 1996). Recent years have seen AMS used extensively in archaeology and in the history of art.

13.5.3.7 Extended X-ray Absorption Fine Structure (EXAFS)

This is a technique that makes use of a monochromator to select and tune the wavelength of synchrotron radiation. The resolution can be as small as 1 eV in a spectrum of 10 keV. As the wavelength of the radiation is scanned, sharp rising edges appear in the absorption spectrum as the energy threshold to excite, for example, the electrons in the K shell of an atom, is reached. The technique is sensitive enough not only to analyse the atomic and molecular constituents in a material but also to deduce interatomic distances. It is extensively used in research into the production of catalysts and the structure of the molecules of biochemistry.

13.5.4 Techniques for revealing the structure of crystals and molecules

13.5.4.1 Diffraction

Revealing the repetitive structure of a crystal by observing the diffraction pattern produced when it is illuminated with monochromatic waves is a basic experimental technique of physics. Much of today's rapid progress in understanding the structure of materials and the composition and shape of the many complicated biochemical molecules that control our bodies' development and health results from diffraction studies (Mutsaers *et al.* 1996).

The more detail we seek, the shorter the wavelength of the probe, whether it is a photon or neutron. Synchrotron light and neutron beams, having wavelengths many orders of magnitude less than visible light, have become powerful probes for this work. Accelerators to produce these probes are either electron synchrotrons of a few GeV dedicated to produce synchrotron radiation, or proton synchrotrons whose intense beams of a few hundred MeV produce neutrons when they hit a metal target—a process referred to as spallation.

The ordered pattern of a crystal can act like a three-dimensional diffraction grating producing a pattern which, when analysed, contains the information necessary to reconstruct the scattering object's shape. If the radiation is strong enough, amorphous samples can also be used. Here one relies on the fact that the few crystals which subtend the Bragg angle between the source and the observer contribute to the pattern, rather as the rain drops make a rainbow. Of course, the random orientation of the target results in circular haloes rather than the clear spots that crystals produce, but these can be disentangled. Molecules, such as proteins and viruses, are studied in this way as are the repetitive structures in polymers and other large molecules. Synchrotron light and neutron beams are used for today's scattering experiments.

As these tasks become more challenging, brighter sources of radiation are needed, but eventually there comes a point where the object to be studied is too small to have repetitive features. It is then that one must turn to X-ray spectroscopy and, in particular, EXAFS described above, by which the interaction of the electron with its surrounding atoms may be revealed. This has been used to study such materials as catalysts and surface layers on industrial glass. Only synchrotron light sources may be used for this. Neutrons do not interact with the electronic structure of atoms.

As just one example of such research techniques applied to the development of new polymers, Boeing replaced aluminium with glass-filled poly-ether–ether–ketone resin developed through synchrotron light research which allowed them to reduce the weight of a Boeing 757 by 30%.

The list of fields of research and application for synchrotron light and neutrons is impressive (Table 13.4).

13.5.5 Synchrotron radiation sources

Electrons circulating at high energy in a synchrotron or storage ring emit a tangential beam of synchrotron radiation over a wide range of frequencies from visible wavelengths into the X-ray region. Many electron rings in the energy range from 1.5 to 8 GeV have been built to serve a number of experiments arranged around their circumference with beams of synchrotron light. One of the largest is the 6 GeV machine, the ESRF at Grenoble. Wiggler magnets and undulators, placed at the point of emergence, enhance the brilliance of the cone of radiation, and monochromators are used to select narrow bands of wavelength where required.

Table 13.4 Fields of structure research

Fields of structure research
Crystal structure with large cells (proteins and enzymes)
Lattice dynamics
Phase transitions
Diffusion in solids
Metal–H_z systems
Interfaces—bonding between semiconductors + insulation
High T_c materials
Magnetic materials
Polymers
Defect structures (stress and fatigue)
Fullerines
Liquids and quantum liquids
Soft matter

The many users who gather around the perimeter of these machines come and go much more frequently than their high-energy physics colleagues. They may be research workers in fields as diverse as the study of the structure of materials such as the hardening of ceramics, or molecular biologists interested in the structure of HIV protein or the SV4D virus, known to induce tumours. This science of designing molecules to modify protein behaviour with new drugs and configure enzymes to promote industrial processes is a rapidly growing field in which experiments using synchrotron light play a crucial role.

Another industrial use of synchrotron light, yet to be fully exploited, is X-ray lithography (Basrour *et al.* 1996). A pattern created on a mask is transferred by X-rays onto a wafer coated with photoresist, which is then developed and the surface etched away, allowing semiconducting circuits to be produced with even greater precision than the conventional UV etching. The nominal present-day precision of 0–5 μm can be reduced by a factor of 5, leading to even faster and compact computer chips. In Fig. 13.11 we see a portable synchrotron light source built for IBM and in Fig. 13.12 the principle of lithography.

13.5.6 Spallation sources

Neutron beams complement synchrotron light as probes for the study of condensed matter and molecular structure. Although comparable in wavelength, the intensity and brightness of neutron sources for scattering studies cannot compare with synchrotron radiation. However, neutron beams penetrate deep into bulk materials and interact principally with nuclei, while the electromagnetic interaction of synchrotron radiation is mainly with atomic electrons. One advantage of neutrons is that their weak interaction results in much less damage in the study of biological material. Another is that when their wavelength is matched to the dimension of the cells of a crystal their energy is comparable to that of the elastic modes within the crystal. This is not the case for synchrotron radiation.

High-flux reactors provide fluxes of neutrons as high as 10^{15} neutrons $cm^{-2}\,s^{-1}$, but at the cost of very high power densities in the reactor core. In a fast neutron

Fig. 13.11 Helios (CERN courier photo).

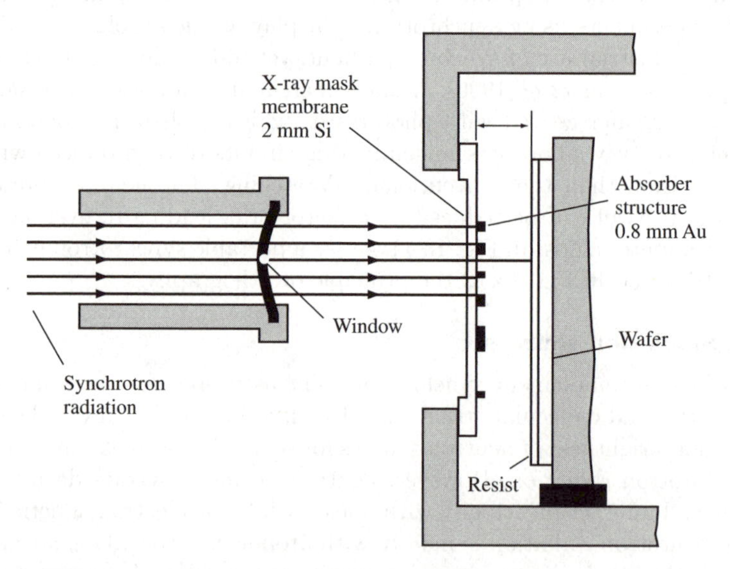

Fig. 13.12 Principle of lithography.

beam from a high-flux reactor, the wavelength is selected with a monochromator, and inevitably the flux is wasted.

Only about one-fifth of the energy is involved when neutrons are produced from a 1 GeV proton from a spallation source. The average number of neutrons produced is about 25 and they must be slowed down with a hydrogen-rich moderator, from which they emerge as a white spectrum.

The beam from a synchrotron is pulsed, typically a short (<1 μs) burst every 20 ms. Over a flight path of 20 m, fast neutrons of wavelengths between 0.1 and 0.4 nm arrive at the detector spread over a 20 ms interval and may be resolved with a time (wavelength) precision of 1/2000. Whatever the accelerator used as a source, the mean current is of paramount importance. This translates to power (between 0.1 and 5 MW) delivered to the target. Spallation sources produce a peak power in the pulse which exceeds that expected from a reactor.

Spallation sources under study hope to raise the power from 160 kW (world record, presently held by ISIS) to 1 MW (ANC, SNS LAMPF, and AUSTRON), and the ambitious ESS in Europe which aims for 5 MW. ESS injects ~600 turns directly from a 1.33 GeV linac into each of two accumulator rings (Lengeler 1990). The beam from each ring is extracted directly after the other to produce the pulse on the target. These projects use H^- linacs with fast stripping injection schemes. In order to inject many turns, vertical orbit bumps and the effect of horizontal displacement are used to fill the phase space in the transverse plane. Dispersion coupled via $\Delta p/p$ fills the longitudinal phase space. SNS will inject 1200 injection turns into a single ring using both vertical and horizontal bumps. The Japanese project JHF aims at 2700 turns.

13.6 Heavy-ion fusion

It is, of course, everyone's dream to produce energy by fusion and, as a first step, to demonstrate 'ignition' at the Lawson criterion

$$n\tau \approx 10^{15}\,\mathrm{cm}^{-3}\,\mathrm{s},$$

where n is the density and τ the confinement time. Firstly magnetic confinement, then lasers, and finally the beams of heavy ions have been considered as a means to achieve this. In the case of ions, as with lasers, first ideas—the so-called direct method—envisaged a large number of beams hitting a tiny pellet of frozen deuterium–tritium 'fuel' from all sides. It turns out that the synchronism and uniformity of the beam distribution required is critical and, in modern 'indirect-drive' methods, ion beams coming from two or more sides convert their energy into X-rays which, within a casing acting as a 'Hohlraum' or black body enclosure, transfer the impact energy uniformly to the pellet (Fig. 13.13) (Hofmann 1996).

The topology of an accelerator complex required to do this is, nevertheless, formidable. In Fig. 13.14 we see a 10 GeV linear heavy-ion accelerator, fed by 16 ion sources of three distinct species of ions which pass through four funnelling stages of RFQs. The 10 GeV ions are stored in 12 storage rings (Prior 1998). Each ring contains 12 bunches which are unloaded and synchronized before acceleration in six induction linacs.

To further improve current concentration, the six induction linacs unload simultaneously from quadrants, the path length to the target being made equal rather like the exhaust manifold of a high-performance racing car. The final

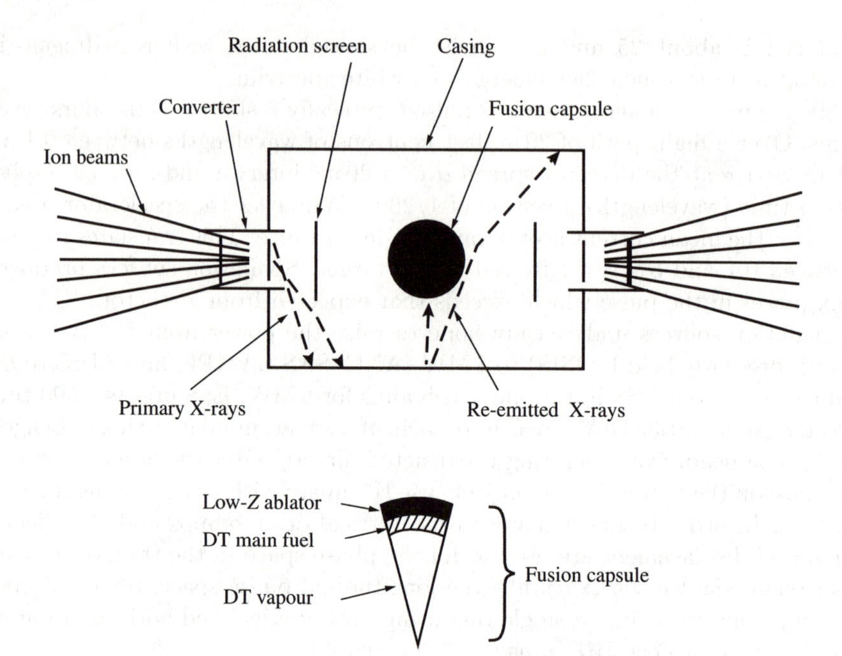

Fig. 13.13 The 'Hohlraum' for a laser-induced fusion experiment.

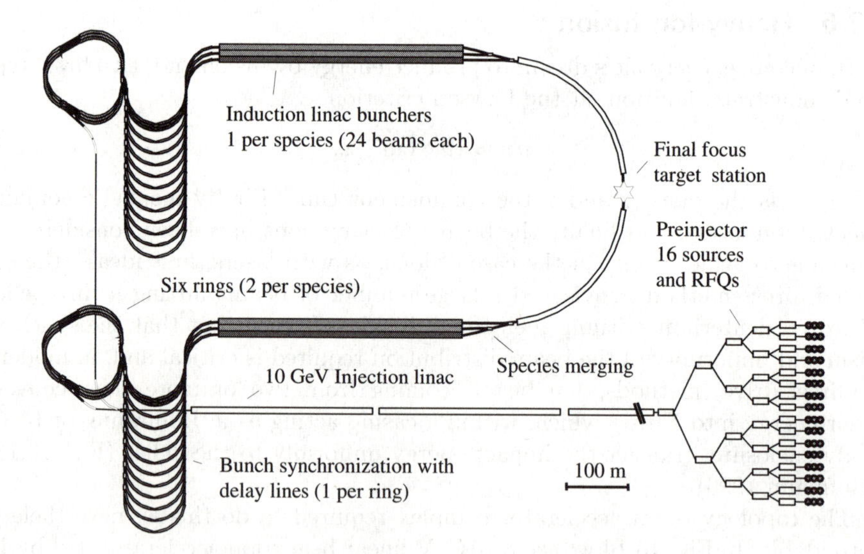

Fig. 13.14 The heavy-ion fusion driver.

energy and the three likely candidates of heavy ions (Bi, Th, and Rh) are selected so that their different masses allow them to catch up at the target. The space-charge forces, which threaten to defocus the beam in the last few metres as it converges on the target, argue for the highest possible energy while the requirement that the ions deposit all their energy within the small thickness of the pellet assembly requires that the energy is low enough to lie high on the steep side of the energy loss versus energy curve. The compromise is much easier to achieve for heavy ions. At each stage, the limits of accelerator technology are challenged.

13.7 Waste transmutation and the energy amplifier

The hope of obtaining limitless electricity from nuclear power has been dampened in the last few decades by the realization that power stations generate a legacy of poisonous actinides with half-lives which are long on a geological scale and that accidents in the reactor and in the waste processing and storage industry are, in the public's perception, rather likely. One of the solutions proposed for dealing with the nuclear waste from power stations is to use a high-current proton beam to convert long-lived nuclides into others which are either stable, short-lived, or which may be used as fuel again. Such proposals have been made by a Los Alamos group, whose high-current LAMPF proton linear accelerator comes closest to the currents required. The viability of the proposal depends upon the cost of the linac and the power required to run it, but it is argued that this will always be much more than the cost of doing nothing and just storing the material underground.

The energy amplifier (Rubbia 1996) offers a means to eliminate this waste, but at the same time produces power on a scale which pays for its capital and running costs (Fig. 13.15). This is an application of accelerators which, if it comes to pass, might affect the prosperity and way of life in the developing world as much as in industrialized countries.

Many of the fears of nuclear power experienced by the general public may be drastically reduced if thorium rather than uranium is used for power production. The waste is a much more benign list of nuclides with lifetimes shorter than the 700 years used as a benchmark. Thorium is plentiful and can be burnt in its natural form requiring no separation. It may be assembled into a reactor which is both stable and sub-critical, depending for its supply of neutrons on a 1 GeV proton beam of about 10 mA current. The term 'energy amplifier' refers to the ratio of output power to that required to operate the accelerator. Clearly, the aim is to make this as large as possible and the challenge is to build a high-current accelerator system which converts as much of its wall-plug power into beam power as possible.

If the ratio of beam power to that drawn from the mains is small, the overall amplification becomes rapidly uneconomical. The choice of accelerator system is evidently crucial. Linacs have been considered but proponents now favour a chain of cyclotrons. The energy required, 1 GeV, is within the range of cyclotrons and, although these accelerators tend to involve a bulky and expensive magnet system, they can run continuously. Modern high-current cyclotrons, PSI, TRIUMF, etc., approach the required current.

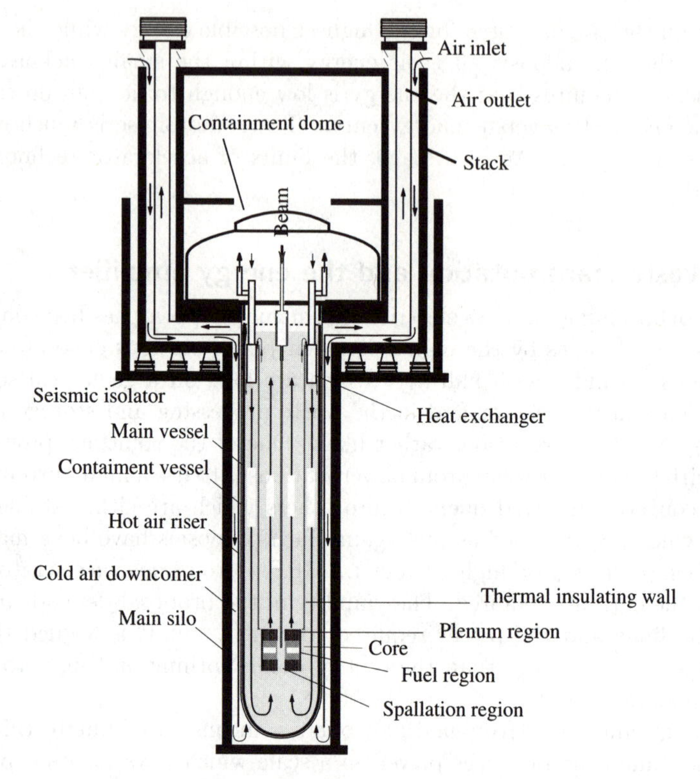

Fig. 13.15 The energy amplifier.

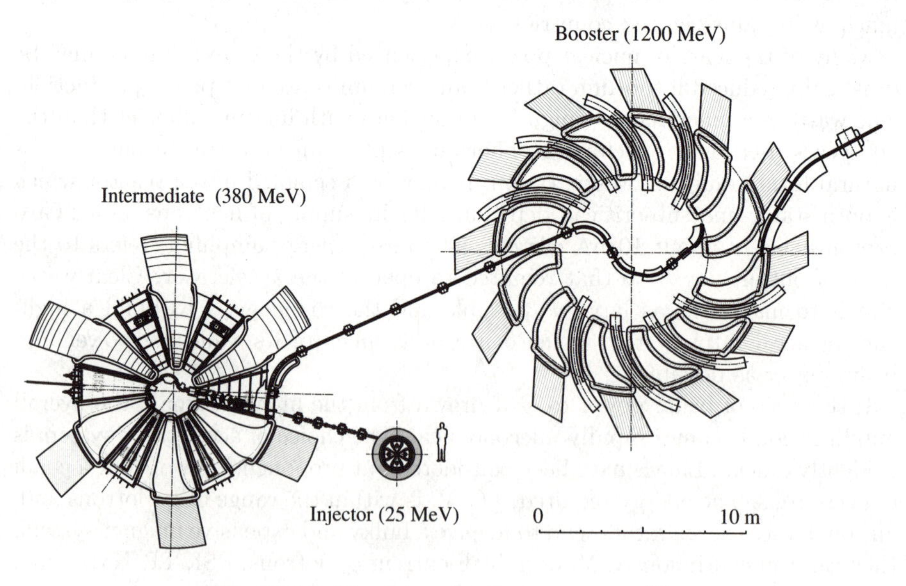

Fig. 13.16 The driver for the energy amplifier.

These modern cyclotron systems (Fig. 13.16) are very different from the simple machines built by Lawrence in the 1930s (Mandrillon *et al.* 1996). Their magnets are split into a number of C-shaped sectors between which there is room to interpose enough r.f. acceleration cavities to keep the turns separated. The entrance and exit faces of the poles of the magnet sectors are curved to provide a constant amount of alternating or 'ridge' focusing per turn. This extra focusing decouples us from the need to have a weaker integrated field at large radii and one can actually increase the field with radius to maintain isochronism as relativity destroys the classical invariance of revolution frequency. Complete turn separation means that the beam may be extracted without destroying the septum deflector which is the essential element in the extraction system. Beam loss is minimal and the transmission efficiency of the accelerator is close to 100%.

Space-charge forces in the transverse and longitudinal planes disturb this focusing and modify bunch length, energy spread, and emittance. For this reason we cascade a chain of cyclotrons, the smaller ones with parameters chosen to combat the space-charge forces. In the scheme currently proposed for the energy amplifier, the first level is a pair of cyclotrons each fed by its own proton source and RFQ. The second level of cyclotron has relatively straight sectors appropriate to its non-relativistic energy range. The last machine with many curved sectors brings the beam to 1.0 GeV.

A pilot project is now about to be launched, intended to stimulate the industry to go into mass production.

Exercises

13.1 An electron beam of 2 MeV and a mean current of 5 mA passes through a 2 mm thick plastic ribbon. The beam width is 15 cm and the ribbon (density 1.4 g/cm^3, $dE/dx = 2.1\,\mathrm{MeV/(g\,cm^{-2})}$) travels at 80 cm/min. Calculate the beam intensity in electrons per second and the total beam power.

13.2 Calculate the area swept per second and the power deposited in the film.

13.3 Calculate the mass of material irradiated per second and the dose received in kGy ($1\,\mathrm{kGy} = 1\,\mathrm{kw\,s\,kg^{-1}}$).

13.4 Synchrotron light of 1 A is a useful probe for molecular structure. Compare its resolving power with the scale of crystal structure, DNA, organic molecules (benzene), simple atoms, and nuclei.

13.5 What energy neutrons give a comparable resolution to a synchrotron light of 1 A?

13.6 Compare a high neutron flux reactor (10^{15} neutrons cm^{-2} s^{-1}) with the intensity of a typical spallation beam of 0.4 MW at 1 GeV protons producing 25 neutrons per proton at a pulse rate of 50 Hz.

14 The future

In Chapter 1 we traced the history of accelerators from Lawrence, van de Graaff, and Wideröe to the present day. We saw how particle physics had stretched the energy scale by many orders of magnitude in the endless quest for tools of greater precision. To achieve this, accelerator builders were quick to exploit the new ideas of storage ring colliders and superconducting magnets and cavities as they emerged, yet the scale of circular machines has inexorably expanded. Indeed, all these machines were based upon concepts invented by the pioneers of the 1930s and 1940s. The technology of lasers, plasmas, and semiconductors made little impact upon accelerators. Although lasers and plasma waves seem to offer much more powerful accelerating fields than conventional or even superconducting cavities, we have not, as yet, found a practical way to harness these fields to provide continuous acceleration. New ideas are still in demand and this chapter is intended to stimulate the quest.

One way to look for new ideas is to identify some general principle in those machines which have proved themselves or shown promise as accelerators and then look for a novel manifestation of that principle. One such principle is that of the 'two-beam' accelerator.

14.1 Two-beam accelerators

In a sense, all accelerators are two-beam devices, since somewhere in the production of the r.f. signals that drive cavities there is usually a beam of electrons. However some are more obviously two-beam devices than others. The CERN linear collider proposal CLIC (Chapter 11) is typical of a two-beam accelerator (Delahaye *et al.* 1999). An intense beam in one linac gives energy to the cavities of another linac that finally does the acceleration to high energy. A number of such schemes have been proposed before alighting on this proposal. In each case the dense low-energy beam acts as a source of energy for a much less intense but higher-energy beam. The simplest of these schemes is the wake-field idea. In this the two beams pass through the same set of cavities and the fields excited by one beam are used to accelerate the other. However, unless the initial driving bunch is specially shaped, the energy gain by a particle in the second beam can be no greater than the loss experienced by each individual in the drive bunch.

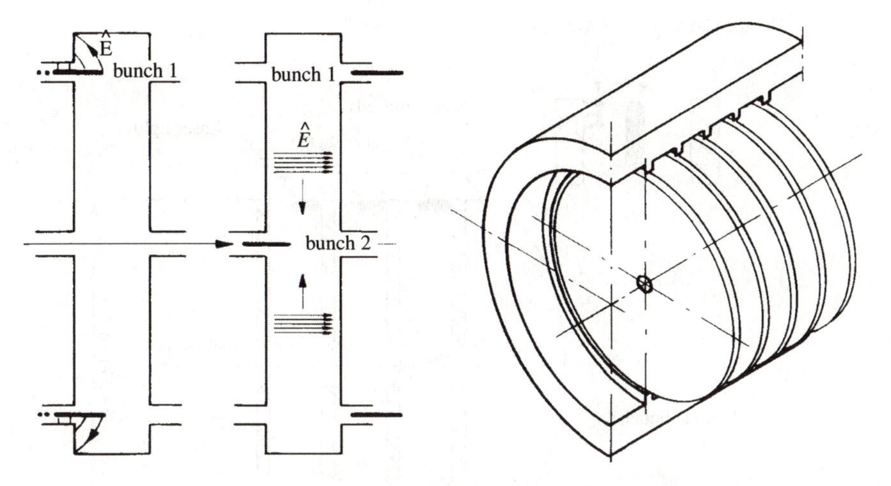

Fig. 14.1 Principle of a wake-field transformer.

Another kind of two-beam device is called the wake-field transformer by its inventors Voss and Weiland (1982) (Fig. 14.1). The driving beam is in the form of a ring accelerated in the direction of its axis through an annular slot surrounding an array of discs. The wake of the ring excites an inward-going wave between the discs. As the diameter of this shock front becomes smaller, the field strength is amplified. When it reaches the small hole at the centre of the discs it increases by a transformer ratio comparable to the radial compression ratio.

In a variation of this device, called the 'switched power linac', the wave is initiated, not by a particle beam but by a ring of laser-triggered spark gaps, whose hold-off energy transforms into an imploding shock wave (Willis 1985) (Fig. 14.2). Of course, this is not strictly a two-beam device.

Finally, and completing this list of devices that have been tried experimentally but have proved rather disappointing, there is the Electron Ring Accelerator (ERA) or 'Smokatron' which attracted much attention in the late 1960s. In this scheme, an electron ring was formed by injecting an intense current at a few MeV into a specially shaped axial field. Ions (protons) resulting from the ionization of background gas collect in the potential well within the ring. The ring is first compressed by increasing the field and then accelerated along its axis in a tapered solenoidal field (Fig. 14.3). Such a field has the property of converting angular momentum into the axial direction. The electron-ring energy in the axial direction is only a few MeV, but protons, trapped in the potential well of the ring, are accelerated with the electrons to the same velocity. The protons, being 2000 times heavier, should then have an energy of many GeV. Earlier many experiments were carried out (Sarantsev and Ivanov 1981) and it was concluded that to accelerate to very high energy would require electron-ring currents beyond the limit of coupled instability between the two species.

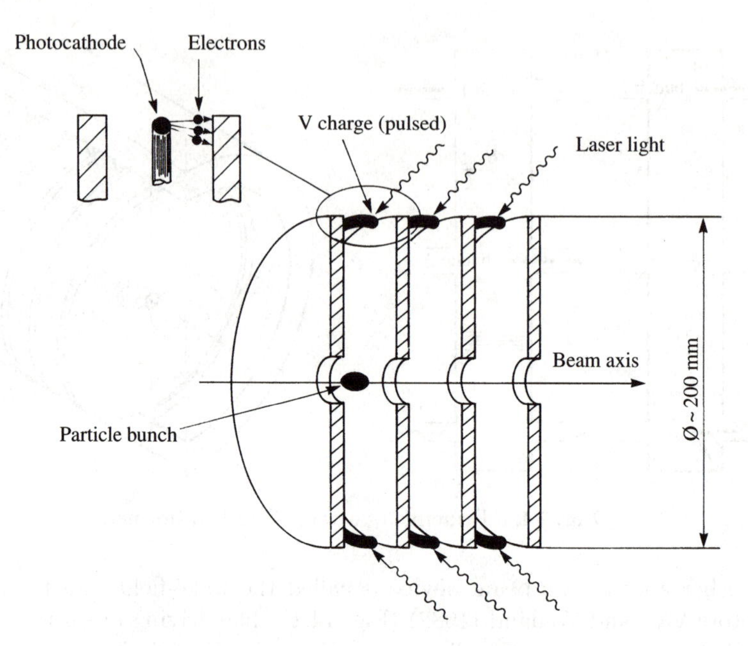

Fig. 14.2 The switched power linac.

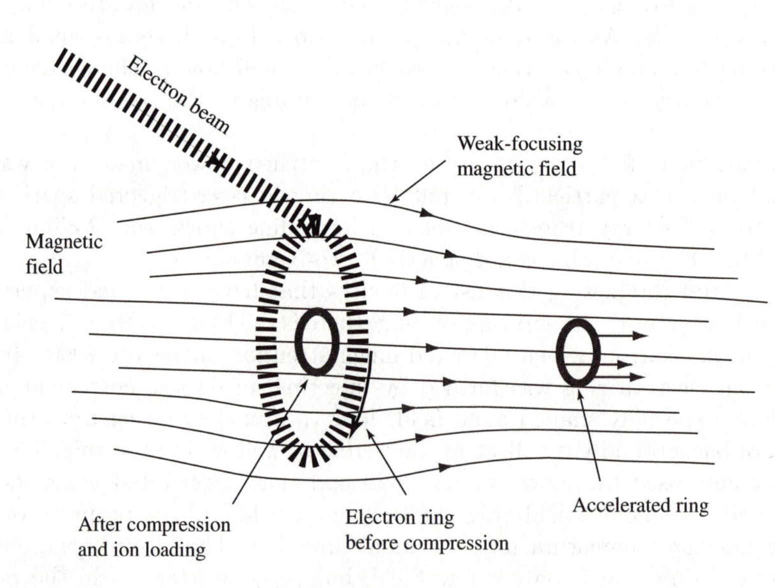

Fig. 14.3 The electron-ring accelerator.

14.2 Direct acceleration with light

14.2.1 In free space

The next approach in our quest for new ideas is to look at a physical system where a high electric field exists and examine under which circumstances it might be harnessed for our use. It is perhaps fortunate that electromagnetic waves do not accelerate particles directly in vacuum or the cosmic-ray flux would be many orders of magnitude more intense and energetic than it is. Life, as we know it, would be impossible unless natural selection had provided us with lead underwear. We saw in Chapter 10 that there are two important reasons for this. The first is that the E vector of plane waves in free space is normal to the direction of propagation. Consequently, the particles are not accelerated in the direction of the wave but merely oscillate transversely in the field of the wave. Secondly, the wave travels with a phase velocity equal to the velocity of light and even if we were to find a way to turn the E vector in the right direction, the particles could not remain in phase with the wave. The wave would change its phase as it overtakes the electron and the net effect would be zero.

There is, of course, a second-order effect. The particle, once accelerated transversely, acquires velocity and becomes a current perpendicular to the transverse B field. The force on this current then deflects this in the direction of propagation of the wave. However, as it is overtaken by the wave, the forces change sign and the result is just a 'quiver' in the form of a figure of 8.

There is a fundamental reason why a massless photon, whose energy $E = \hbar\omega$ is just equal to $pc = \hbar kc$, cannot give all its energy and momentum to a massive particle, say an electron, whose energy is $m_0c^2 + \hbar\omega$. (Note ω and k are the angular frequency and wavenumber of the particle in wave mechanics.) If we normalize k by multiplying by the Compton wavelength, $\lambda = \hbar/m_0c$, we obtain trajectories for the electron and the photon on a (E, pc) or $(\omega,\ k)$ diagram as shown in Fig. 14.4. Readers will recognize this diagram as the dispersion diagram in a plain waveguide we found in Chapter 10 and perhaps remember that its local slope is the group velocity, that is, the velocity of the particle in wave mechanics.

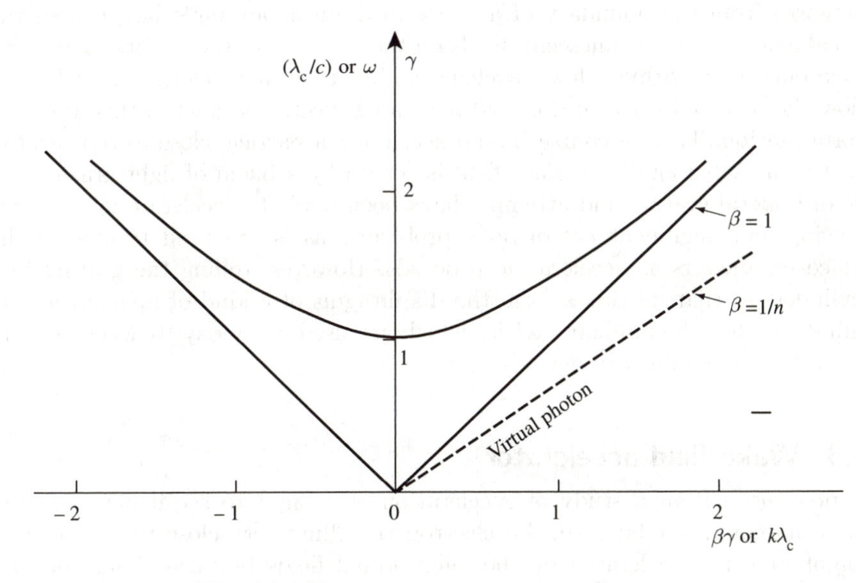

Fig. 14.4 Energy versus momentum diagram.

The photon has velocity c, while the electron velocity only rises asymptotically towards that of the photon. The two curves cannot match each other in slope, yet energy and momentum conservation demands that the increments ΔE and $\Delta(pc)$, and hence the local slope, must match if the photon is to interact with the electron.

There are two ways around this, and these may be the signposts towards a new acceleration principle. One is to allow the photon to recoil and keep some of its energy but with a momentum in the reverse direction. The other is to slow down the photon as it might be, for example, in a refractive medium. The former case is just Compton scattering, a well-known effect, but with a cross section which is too small to effect an efficient transfer of energy, say, from a laser to a bunch of particles. The latter has the disadvantage that other interactions with the refractive medium slow the particle down and scatter it. As we examine other methods of acceleration in use or proposed, we shall find they fall into either one of these two categories and always seem to involve photons whose energy is not equal to momentum (times c), the so-called 'virtual photons'.

14.2.2 Inverse Čerenkov accelerator

This is an example of a virtual photon principle. A particle travelling through a medium of refractive index n will emit a cone of light if its velocity is greater than c/n. The cone angle is $\sin^{-1}(c/n)$. Conversely, a cone of laser light directed on a particle by an axial prism (axicon) should accelerate it. This has been shown to work over a few keV.

14.2.3 Surface waves

It is perhaps not well known, but a beam of light reflected internally as it tries to emerge from the boundary of a dense medium at an angle larger than the critical angle sets up 'evanescent' fields on the rare side of the surface. These fall off exponentially within a few wavelengths, but they move along the surface to follow the reflected wave ($v < c$) and have an E vector parallel to this direction of propagation. This, of course, would accelerate a particle close to the surface. Exactly the same kind of surface field is set up by a beam of light grazing the face of a metal grating and attempts have been made to accelerate in this way. Focusing and alignment are obvious problems, as is the need to change the surface parameters as acceleration proceeds. However, rolling the grating into a cylinder, so that its ridges form the diaphragms of a kind of linac structure, reminds us that the familiar cavities which are used every day to accelerate are part of the same family of devices.

14.3 Wake-field accelerator

We now move from a study of acceleration with light to see if fields excited in plasma could not be used. An electron travelling very close to the velocity of light may be accelerated by the longitudinal fields between charge density waves which can be stimulated in the electrons of a plasma. In the case of the wake-field accelerator, the waves are driven by the wake of a very short and dense bunch of electrons which precedes the electrons we wish to accelerate. This driving bunch can have much lower energy than the electrons which follow. There must be many such stages of acceleration following each other, rather like the tanks of a conventional linac. To understand the principle, we must first write down the fluid equations for a plasma, and we consider the one-dimensional non-relativistic case:

$$\frac{\partial n}{\partial t} + \nabla \cdot (n\boldsymbol{v}) = 0,$$

$$\frac{\partial \boldsymbol{v}}{\partial t} + (\boldsymbol{v} \cdot \nabla)\boldsymbol{v} = \frac{e}{m}(\boldsymbol{E} + \boldsymbol{v} \wedge \boldsymbol{B}).$$

Consider a plasma with density n_0 and an injected bunch with density n_b, and retain only first-order perturbations in $\boldsymbol{v}$, $\boldsymbol{E}$, and $\boldsymbol{B}$. One finds (Ruth *et al.* 1984) that

$$\frac{\partial n_1}{\partial t} + n_0(\nabla \cdot \boldsymbol{v}) = 0,$$

$$\frac{\partial \boldsymbol{v}}{\partial t} = \frac{e\boldsymbol{E}}{m}.$$

The wave equation for the plasma may then be derived:

$$\frac{\partial^2 n_1}{\partial t^2} + \omega_\mathrm{p}^2 n_1 = 0.$$

In this equation, ω_p is the plasma frequency defined by

$$\omega_\mathrm{p}^2 = \frac{e^2 n_0}{m_0 \varepsilon_0}.$$

Consider next a driving beam which is a delta function of charge in the direction of motion, z, and whose local density is then:

$$n_\mathrm{b} = \sigma\delta(z - v_\mathrm{b}t) = \sigma\delta(y),$$

where $y = z - v_\mathrm{b}t$ is a coordinate moving with the bunch and σ the total charge. Now the field is not simply

$$\nabla \cdot E = \frac{en_0}{\varepsilon_0},$$

where n is the plasma density of electrons, but instead

$$\nabla \cdot E = \frac{e(n_0 + n_\mathrm{b})}{\varepsilon_0}.$$

The wave equation for the plasma is then driven:

$$\frac{\partial^2 n_1}{\partial t^2} + \omega_\mathrm{p}^2(n_1) = -\omega_\mathrm{p}^2 n_\mathrm{b}.$$

Clearly, this is an oscillator driven by the wavenumber $k = 2\pi/\lambda = \omega_p/v_b$. We may rewrite this in terms of the independent coordinate y:

$$\frac{\partial^2 n}{\partial y^2} + k^2 n = -k^2 \sigma \delta(y).$$

It turns out that a simple conservation of energy argument suggests that the maximum energy that can be gained by the electron following in the wake of a single symmetric bunch is just twice the energy that each of the electrons of the drive bunch possesses. But, apparently, it is possible to improve on this in the wake of a drive bunch with an asymmetric linear rising edge and a steep fall (van der Meer 1985; Dawson and Chen 1985).

14.4 Beat-wave accelerator

The principles of the wake-field accelerator and the beat-wave accelerator are similar, but in the beat-wave accelerator the plasma waves are driven by electromagnetic waves and not by a particle bunch (Fig. 14.5). Normally, it is impossible to penetrate a plasma with an electromagnetic wave below the plasma frequency, and above the plasma frequency the wavelength of the plasma wave and its phase velocity would not be matched either to the plasma we wish to drive or to the particles to be accelerated. This can be overcome by the laser plasma accelerator scheme or the beat wave scheme both proposed by Tajima and Dawson (1979).

In the beat wave solution we generate a longitudinal plasma wave from the beat frequency between two high-frequency lasers (or two modes of the same laser) each of which is high enough in frequency to penetrate. The difference in frequency is made equal to the plasma frequency so that beats between the wave drive the plasma into strong oscillation (Tajima and Dawson 1982). The relation

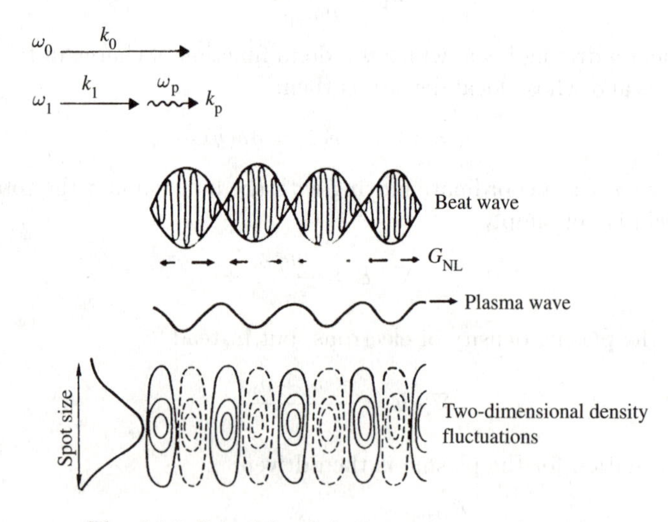

Fig. 14.5 Principle of the beat-wave accelerator.

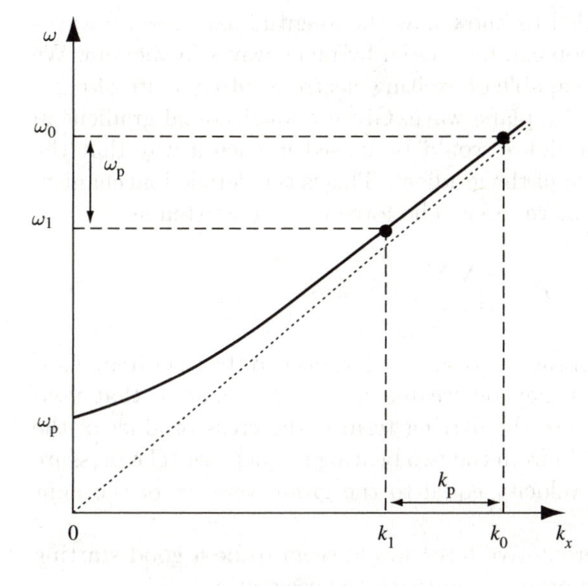

Fig. 14.6 Dispersion diagram for a plasma.

between wavenumber and frequency for a plasma follows the by now familiar curve shown in Fig. 14.6. The slope of this graph is the group velocity of the wave, which can be matched to the phase velocity of the longitudinal electron plasma waves by choosing a plasma density and hence plasma frequency.

The group velocity is given by

$$v_{\mathrm{g}} = \frac{d\omega}{dk} = c\left(1 - \frac{\omega_{\mathrm{p}}^2}{\omega^2}\right)^{1/2}$$

which is just less than c for an underdense plasma $\omega \gg \omega_{\mathrm{p}}$.

The wave and the electrons only have to stay reasonably close in phase over the length of one stage of the accelerator.

The wave equation for the plasma in the beat-wave accelerator is the same as for that in the wake-field accelerator:

$$\frac{\partial^2 n}{\partial t^2} + \omega_{\mathrm{p}}^2 n = 0,$$

and the solution has the form

$$n = \hat{n} \cos k(z - v_{\mathrm{p}} t)$$

with $\omega_{\mathrm{p}} \approx k_{\mathrm{p}} c$. Since $\nabla \cdot \boldsymbol{E} = -en$, the maximum accelerating field is $kE = -en$, which we may write as

$$E = \frac{-m\omega_{\mathrm{p}} c}{e} \approx \sqrt{10 n_0}\,\mathrm{V m^{-1}}.$$

In the laser plasma wakefield scheme short pulse of laser light appearing as a delta function on the scale of the waves to be excited will produce the Langmuir wave.

The reader is probably puzzled to know how the longitudinal pressure waves in the plasma electron population can be excited by plane waves in vacuum. We remember that a plane wave is capable of exciting electrons into a figure-of-eight motion—a second-order effect of a plane wave. Given a longitudinal gradient in the amplitude of the wave, the figure would be biased in such a way that the particle migrates in the direction of the gradient. This is the detailed mechanism of what is called the ponderomotive force. The force may be written as

$$\boldsymbol{F} = \frac{-\nabla \boldsymbol{E}^2}{16\pi n_{\mathrm{ec}}},$$

where n_{ec} is the cutoff density of the electrons with respect to the laser frequency.

The behaviour of the plasma may be treated in a way similar to that used in the wake-field accelerator. Here the driving term is the cross product of the transverse velocity and the B field from the two beating frequencies. The pressure waves produced have a phase velocity equal to the group velocity of the light through the plasma.

Contemplation of the ponderomotive force would seem to be a good starting point for anyone looking for an original approach to acceleration.

14.5 Near fields and virtual photons

Now that we have looked at a few of the possible acceleration mechanisms, it is time to reflect on the general trend. This may give us some idea of where to look and where not. Apart from the need to bend the $\boldsymbol{E}$ vector in the direction of acceleration, there seems to be a fundamental difficulty in accelerating with the plane waves in vacuum. We may associate this with the zero mass of the photon, which fixes the phase velocity of c. Reflecting waves between metallic walls will result in a component of E in the desired direction and, if the walls have ridges, will slow down the phase velocity. A dielectric layer will do the same. In each case, the photon is 'virtual'. Such photons can only exist close to moving charges such as those in conducting walls and in the atoms of dielectrics. If we closely examine the argument for the equality of energy and momentum for a photon (Lawson 1970), we can trace it back to the right-hand side of Maxwell's wave equation being zero:

$$\nabla^2 A - \frac{1}{c^2}\frac{\partial^2 A}{\partial t^2} = -4\pi j.$$

Near a current j, the photon need not have equal energy and momentum. An extreme case, of course, is a constant magnetic field.

We see an echo of this argument—now in the classical context—when we examine the retarded field resulting from a moving charge. The primed quantities refer to the position occupied by the charge when it sent the signal now being observed at a distance r. The unit vector $\boldsymbol{e}$ points to the observer from the charge. Feynmann (1963) has shown that

$$\boldsymbol{E} = -\frac{q}{4\pi\varepsilon_0}\left[\frac{\boldsymbol{e}_{r'}}{r'^2} + \frac{r'}{c}\frac{d}{dt}\left(\frac{\boldsymbol{e}_{r'}}{r'^2}\right) + \frac{1}{c^2}\frac{d^2\boldsymbol{e}_{r'}}{dt^2}\right].$$

Only the last of the three terms in this equation describes a plane wave. The first two are 'near-field' terms, which fall off with the square of the distance and can be thought of as fields in the neighbourhood of conductors or moving charges. The other characteristic of the fields that accelerate particles would seem to be that the $\boldsymbol{E}$ and $\boldsymbol{H}$ vectors are in quadrature and the wave impedance is, therefore, reactive.

An exact equivalence between the different aspects of waves that accelerate is not easy to prove but, provided we do not close our minds to exceptions and paradoxes, near and virtual fields offer more chance of success in a search for a new accelerating method. We move on to look at methods of acceleration which might be at work in interstellar space.

14.6 Astrophysics

Although our quest has revealed a number of possible but seemingly impractical mechanisms, there must be some simple mechanism, as yet undiscovered, which accounts for the presence of very high cosmic rays ($>10^{20}$ eV). It is surprising that accelerator builders, their eyes no doubt directed earthward, take little heed of the very high energy acceleration mechanisms in space. Perhaps, this is because even astrophysicists are hard-pressed to explain the origins of cosmic rays.

The original explanation of Fermi suggested that rapidly moving clouds of interstellar matter accelerate electrons, and this was extended to include two approaching clouds between which the electrons would bounce many times, gaining energy each time (Fig. 14.7).

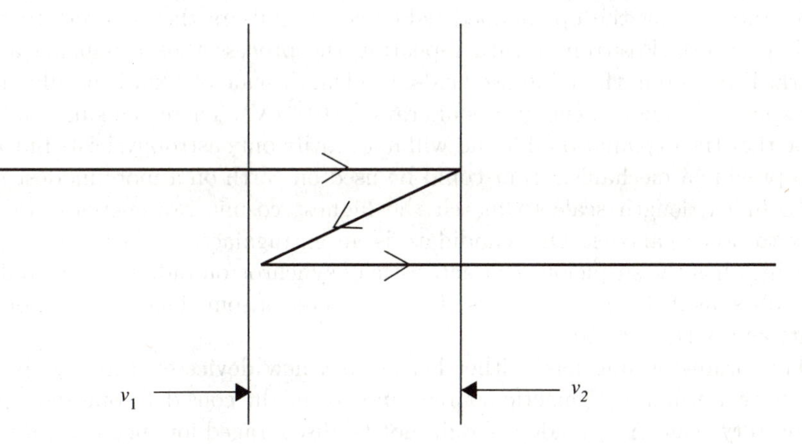

Fig. 14.7 Fermi's explanation of acceleration by colliding waves of matter.

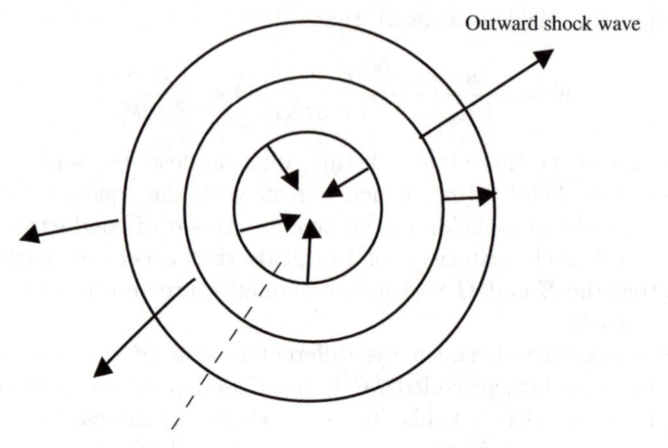

Fig. 14.8 Supernova shock-wave model.

The present thinking of astrophysicists is that, during the collapse of the stellar core of a supernova, outward shock waves are produced (Fig. 14.8). The model, which assumes a reasonable r^{-1} variation of density, predicts that the γ of the material would follow

$$\gamma \propto \rho^{-1/2(3-\sqrt{3})},$$

which would, in turn, explain why the energy spectrum of cosmic rays falls as $E^{1.6}$.

On the other hand, it may be that the intense 10^8 T magnetic fields around pulsars play some role (Fig. 14.9). It has been suggested that as electrons are accelerated along the open-ended field lines of pulsars they convert to γ-rays and reconvert. Positrons return repeating the process thus producing a giant spark. But even in these intense fields, the length scale of 1 km is hardly enough to explain the highest-energy cosmic rays ($>10^{20}$ eV). There is a small but faint hope that the explanation if found will not gratify only astrophysicists but might also provide a mechanism that could be used on earth on a more modest scale.

To find a length scale to match the highest cosmic ray energies, we must look to other galaxies. One candidate is an extragalactic radio source Cygnus A, which has the suspicious characteristic of synchrotron radiation emitted from two lobes, as if the opposite ends of the diameter of some huge accelerator, light years across (Fig. 14.10).

This brainstorming has neither led us to a new device on a terrestrial scale nor have we found a galactic source, and we might consider compressing to a laboratory scale. Yet, readers should not be discouraged for such exercises serve to stimulate pathways in the inventive brain which will one day bear fruit.

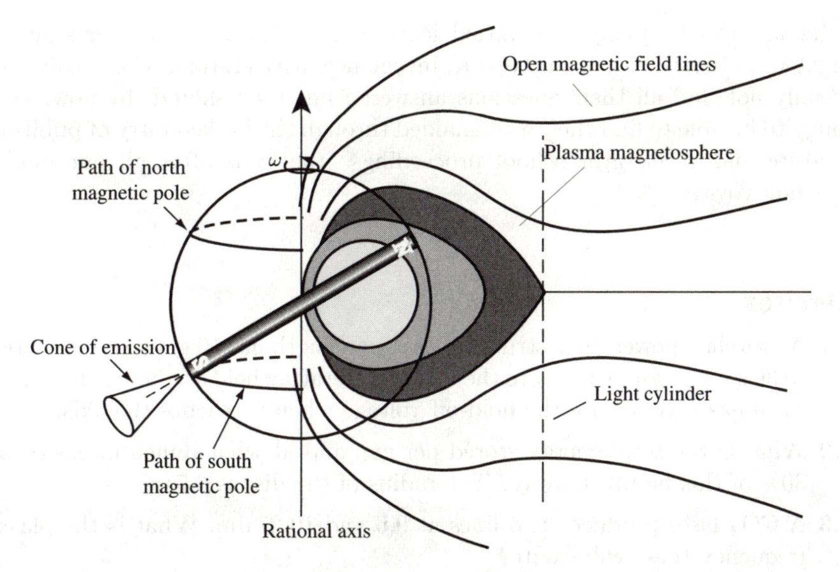

Fig. 14.9 Pulsar.

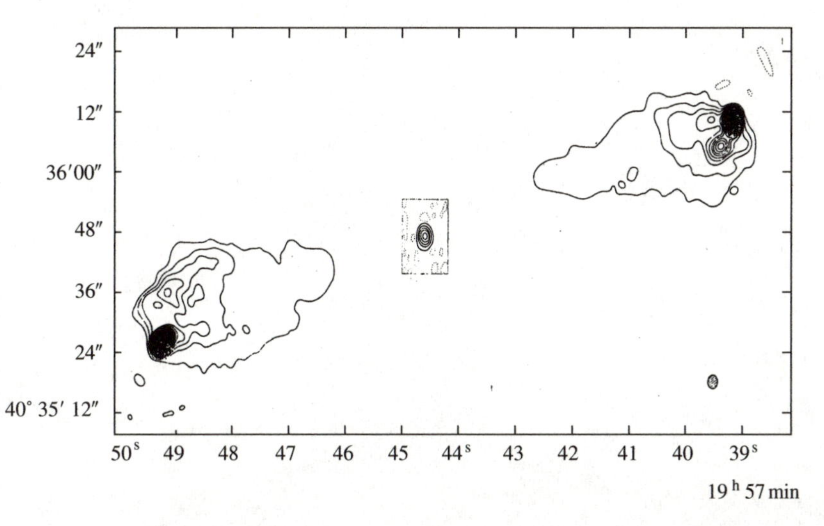

Fig. 14.10 Radio emission from an active galaxy.

14.7 Conclusions

We have travelled a long road from Wideröe musing whether a circulating beam could be the secondary of a transformer, to our own speculation concerning the electromagnetic 'engines' which accelerate cosmic rays. Along the way, we have admired the existing landscape of accelerator technology. Hopefully, our excursions into its many and varied features have given the reader sufficient insight to understand and perhaps to invent new accelerators. The readers will certainly not find all their questions answered but they should, by now, know enough to be able to find their way unaided through the backcountry of published literature and accelerator school proceedings—that was after all our modest aim—bon voyage.

Exercises

14.1 A switched power linac structure of 50 cm length has 10 gaps and is charged to 30 kV per gap. Calculate the total accelerating field assuming the voltage wave has seven times the hold-off voltage when it reaches the axis.

14.2 What is the total energy stored per gap and at what bunch intensity will 30% of this be taken away? The radius of the disc is 0.7 m.

14.3 A CO_2 laser produces two lines at 9.6 and 10.27 μm. What is the plasma frequency these will excite?

14.4 What density of plasma will be excited by this beat frequency?

Appendix: Synchrotron radiation

A1.1 Retarded fields

One might think that it would be simple to apply elementary electromagnetic theory in order to understand the waves set up by a moving charge. However, it is really quite complicated because at the heart of the treatment is the fact that although the electron and observer are travelling slowly, the radiation, travelling with velocity c, takes a certain time, $t = s/c$, to reach the observer and during this time the charge has changed its position and velocity (see Fig. A1). The field at the observer at any time t is, therefore, determined by the position and velocity, not at time t but at some earlier time t'. We call the position, velocity, and the earlier time, the 'retarded' coordinates of the charge.

A1.2 Solution to Maxwell's equations

When we write down the solution to Maxwell's equations for the magnetic vector potential at a point 1 due to a charge distribution at another point 2, it should have the form

$$\mathbf{A}(1, t) = \int \frac{\mathbf{j}(2, t - r_{12}/c)}{4\pi\varepsilon_0 r_{12} c^2}\, dv;$$

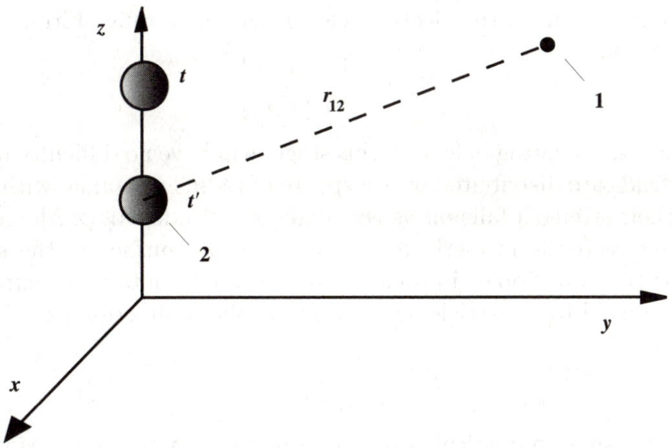

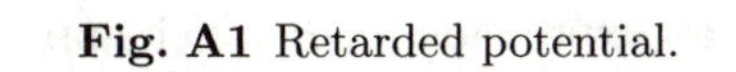

Fig. A1 Retarded potential.

here r_{12} is the line joining the observer and the position of the charge at the time t' at which it emitted the signal, which is then observed at point 1 at time t.

Now let us take a single blob of charge q, displaced in the z-direction and with mean position $\bar{z}$. Its dipole moment is defined as $\bar{p} = q\bar{z}$ and we can substitute it for $\mathbf{j}$ in the above expression:

$$\int j_z\, dv = \dot{\bar{p}} = q\dot{\bar{z}},$$

and if we are prepared to believe that all the parts of the blob of charge are at approximately the same distance r from the observer, we may write

$$\mathbf{A}(t) = \frac{1}{4\pi\varepsilon_0 c^2}\frac{\dot{p}}{r}.$$

At this point we should not lose sight of the fact that it is the retarded value of $\dot{p} = \dot{p}(t - r/c)$ that we must use and that this has spatial derivatives because it is a function of r. Thankfully, for the particular geometry chosen, we can assume that $\partial/\partial y = \partial/\partial r$. Differentiating the above expression to obtain the magnetic field at the observer, we obtain two terms (within rectangular brackets):

$$B_x = \frac{1}{4\pi\varepsilon_0 c^2}\frac{\partial}{\partial r}\left(\frac{\dot{p}}{r}\right) = \frac{1}{4\pi\varepsilon_0 c^2}\left[\dot{p}\frac{\partial}{\partial y}\left(\frac{1}{r}\right) + \frac{1}{r}\frac{\partial}{\partial y}(\dot{p}(t - r/c))\right],$$

$$B_x = \frac{1}{4\pi\varepsilon_0 c^2}\left[-\frac{\dot{p}}{r^2} - \frac{1}{r}\frac{\ddot{p}}{c}\right] = \frac{1}{4\pi\varepsilon_0 c^2}\left[-\frac{q\dot{z}}{r^2} - \frac{1}{r}\frac{q\ddot{z}}{c}\right].$$

The first term falls off with the square of the distance while the second as $1/r$. The first is the static field while the second, a direct consequence of putting in the relativistic retarded position, is the field radiated to more distant observers. Note that the radiated field is proportional to *acceleration* of the charge, but not the current.

A similar treatment of the electric field yields a long-range term:

$$E_y = \frac{1}{4\pi\varepsilon_0}\left[\frac{1}{r}\frac{q\ddot{z}}{c}\right].$$

Most readers, accepting defeat by this stage, will have no difficulty in believing that these fields are distributed over a sphere of radius r co-axial with the dipole such that their strength falls off as the angle of latitude, $\cos\varphi$. Moreover, when the Poynting vector is properly integrated over the surface of the sphere and summed for the two dipoles in quadrature, we would obtain the expression for the power emitted by a particle accelerating in the y-direction as

$$P = \frac{1}{6\pi\varepsilon_0}\frac{e^2}{c^3}\cdot(\ddot{z})^2.$$

This, but for a small numerical correction due to our integration, is the Larmor expression. Now we have an expression for P which applies in the moving system of the electron and we can, if we choose, write f for the acceleration in the z-direction.

A1.3 Lorentz transformation

Now suppose both the electron and an observer are moving together around the circumference of a synchrotron; then the principle of special relativity suggests that the values of acceleration and power seen by the observer must fit the above expression. However, we are seeking an expression which is valid for an observer at rest in the laboratory. The first point to remember is that the power emitted is a Lorentz invariant.

To be convinced of this, it helps to think of the photons emitted by the accelerating or decelerating electron. Because a massless photon has identical energy and momentum, conservation of the electron's energy and momentum can only be satisfied if two photons are emitted in different directions, so that their energies add but their momenta subtract.

In the moving frame in which the electron has no momentum, the two photons must have the same energy and be emitted back to back. In the laboratory frame, the forward photon is blue-shifted to a higher energy and the backward photon red-shifted to a lower energy. A little consideration of the application of the Lorentz transformation should convince the reader that the sum of the energies of the two photons increases with γ but emission events occurring at a constant rate in the moving system appear, in the laboratory, to be separated by intervals which are also increased by γ. These two effects cancel, the power taken away in the moving frame is the same as in the laboratory and, hence is Lorentz invariant.

Another consequence of the transformation into the laboratory frame is that an isotropic distribution of photon emission in the rest system of the electron is observed in the laboratory as a narrow cone in the forward direction with an opening angle $1/\gamma$. Hence if we calculate P in the moving system, it should have the same value as in the laboratory. This is not the case if we just use the Larmor expression because $(\ddot{z})^2$ is not invariant under Lorentz transformation, and suggests that we must generalize the above expression to its Lorentz invariant form. In this form it should give the same value of P whether we express the acceleration in the laboratory or in the moving frame. The Lorentz transformation for an acceleration *transverse* to the relative velocity vector of the two systems conserves the invariant $f\gamma^2$ and, thus if we replace $(\ddot{z})^2$ by $f^2\gamma^4$ we leave the expression unchanged in the moving system where γ is unity. At the same time it ensures that P is indeed the same if we apply the formula in the laboratory frame:

$$P = \frac{1}{6\pi\varepsilon_0}\frac{e^2 f^2}{c^3} \cdot \gamma^4.$$

Now we see the strong energy dependence of the radiated power appearing in this expression.

Answers to exercises

1.1 $E = E_0 + T, \quad E^2 = E_0^2 + (pc)^2, \quad pc = \sqrt{(E_0 + T)^2 - E_0^2}$

1.2 $E_0 = m_0 c^2, \quad E = \dfrac{m_0 c^2}{\sqrt{1-\beta^2}}, \quad p = \dfrac{m_0 v}{\sqrt{1-\beta^2}},$

$$\frac{E}{E_0} = \frac{E}{m_0 c^2} = \frac{m_0 c^2}{m_0 c^2 \sqrt{1-\beta^2}} = \frac{1}{\sqrt{1-\beta^2}} = \gamma,$$

$$\frac{pc}{E} = \frac{m_0 v/\sqrt{1-\beta^2}}{m_0 c^2/\sqrt{1-\beta^2}} c = \frac{v}{c} = \beta.$$

1.3
$$\begin{aligned} E &= T + m_0 c^2 \\ &= 1 + 0.9383 = 1.9383\,\text{GeV}. \end{aligned}$$

1.4
$$1.9383^2 = (0.9383)^2 + (pc)^2,$$
$$p = (1/c)\sqrt{1.9393^2 - 0.9383^2} = 2.8766\,\text{GeV/c}.$$

1.5 The field rises to 1 T in 5 ms. The flux linking the orbit is $2 \times \pi\rho^2 B$ and its rate of rise is

$$2 \times \pi \times (0.1)^2 \times \frac{1}{5 \times 10^{-3}} = 12.6.$$

This rate of change of flux is just the voltage per turn ($\int E\,ds = \int\int (dB/dt) d\tau$) i.e. 12.6 eV/turn. The number of turns is

$$5 \times 10^{-3} \times \frac{c}{2\pi\rho} = 2.385 \times 10^6,$$

and hence the top energy is this number times 12.6, that is, 30 MeV.

1.6 The balance of centripetal force and centrifugal acceleration gives

$$m\omega^2\rho = evB.$$

The angular revolution frequency is $\omega = v/\rho$. Thus, $\omega = (e/m)B$. At 1.2 T

$$f_{\text{rev}} = \frac{\omega}{2\pi} = \frac{9.58 \times 10^7}{2\pi} \times 1.2 = 18\,\text{MHz}.$$

1.7 The quantity known as magnetic rigidity is defined in Section 2.2:

$$B\rho\ (\text{T m}) = 3.3356 pc\ (\text{GeV}).$$

For deuterons the argument of Section 2.2 is unchanged and, therefore, the maximum momentum accelerated will be the same as for protons:

$$\begin{aligned}
c^2p^2 &= E^2 - E_0^2,\\
T &= 1.0\,\text{GeV},\\
c^2p^2 &= E^2 - E_0^2,\\
c^2p^2 &= (1.0 + 0.938)^2 - 0.938^2 = 2.875,\\
cp &= 1.696\,\text{GeV},\\
c^2(p_{\text{deu}})^2 &= (T_{\text{deu}} + 2 \times 0.938)^2 - (2 \times 0.938)^2 = 2.875,\\
T_{\text{deu}} &= 0.653 \text{ or } 0.326\,\text{GeV per nucleon.}
\end{aligned}$$

1.8 The answer to Exercise 1.1 gives

$$\beta = \frac{cp}{E}.$$

Calculating the revolution frequency for protons:

$$\begin{aligned}
\beta_{\text{prot}} &= \frac{1.696}{1.938} = 0.875,\\
f_{\text{rev}} &= \frac{\beta c}{2\pi R} = \frac{0.875 \times 2.9979 \times 10^8}{2\pi \times 25} = 1.67\,\text{MHz};
\end{aligned}$$

and repeating for deuterons:

$$\begin{aligned}
\beta_{\text{deu}} &= \frac{1.696}{0.653 + 2 \times 0.938} = 0.670,\\
f_{\text{rev}} &= \frac{\beta c}{2\pi R} = \frac{0.670 \times 2.9979 \times 10^8}{2\pi \times 25} = 1.28\,\text{MHz}.
\end{aligned}$$

2.1 64.5 m.

2.2 26.68 and 0.4136 T.

2.3 12 T m^{-1}.

2.4 0.509 m.

2.5 4.035 m.

2.6 96 and 48, which fit in with a superperiodicity of 6.

2.7 61.7 m and 41%.

3.1 Differentiate twice:

$$\begin{aligned}
y &= a\beta^{1/2}\cos\phi \quad (\phi_0 = 0),\\
y' &= \tfrac{1}{2}\beta^{-1/2}a\cos\phi \times \beta' - \beta^{-1/2}a\sin\phi\\
&= a\beta^{-1/2}\left(\tfrac{1}{2}\beta'\cos\phi - \sin\phi\right) \quad (\phi' = \beta^{-1}),
\end{aligned}$$

$$y'' = -\frac{1}{2}a\beta^{-3/2}\beta'\left(\frac{1}{2}\beta'\cos\phi - \sin\phi\right)$$
$$+ a\beta^{-1/2}\left(\frac{1}{2}\beta''\cos\phi - \frac{1}{2}\frac{\beta'}{\beta}\sin\phi - \frac{1}{\beta}\cos\phi\right).$$

Substitute in Hill's equation

$$y'' + Ky = 0 = a\beta^{-1/2}\cos\phi\left[\frac{1}{2}\beta'' - \frac{1}{4\beta}\beta'^2 + K\beta - \frac{1}{\beta}\right];$$

therefore,

$$\tfrac{1}{2}\beta\beta'' - \tfrac{1}{4}\beta'^2 + K\beta^2 = 1.$$

3.2 $$M = \begin{pmatrix} 1 & 0 \\ -1/f_2 & 1 \end{pmatrix}\begin{pmatrix} 1 & l \\ 0 & 1 \end{pmatrix}\begin{pmatrix} 1 & 0 \\ -1/f_1 & 1 \end{pmatrix}.$$

3.3 $$M_{\mathrm{FODO}} = \begin{pmatrix} 1 + l/2f & l \\ -l/4f^2 & 1 - l/2f \end{pmatrix}\begin{pmatrix} 1 - l/2f & l \\ -l/4f^2 & 1 + l/2f \end{pmatrix}$$
$$= \begin{pmatrix} 1 - l^2/2f^2 & 2l(1 + l/2f) \\ -(l/2f^2)(1 - l/2f) & 1 - l^2/2f^2 \end{pmatrix}.$$

3.4 97.04°, 23.1 mm.

3.5 Reversing the signs of focal length,

$$M_{\mathrm{DOFO}} = \begin{pmatrix} 1 - l^2/2f^2 & 2l(1 - l/2f) \\ -(l/2f^2)(1 + l/2f) & 1 - l^2/2f^2 \end{pmatrix}$$

Trace/2 $= \cos\mu = 1 - l^2/2f^2$ as before, and $m_{12} = 2l(1 - l/2f)$. Hence,

$$\beta = \frac{2l(1 - l/2f)}{\sin\mu},$$
$$\beta = 3.31\,\mathrm{m}$$

3.6 $$Q = 6.47, \qquad 2\pi R = 6.545 \times 2 \times 24, \qquad R = 50\,\mathrm{m}.$$
$$\beta = \frac{50}{6.47} = 7.73\,\mathrm{m}.$$

3.7 $$\cos\mu = 1 - \frac{l^2}{2f^2}$$
$$d(\cos\mu) = \sin\mu\, d\mu = \frac{l^2}{f^2}\frac{df}{f}$$
$$dQ = \frac{d\mu}{2\pi} = \frac{1}{2\pi\sin\mu}\frac{l^2}{f^2}\frac{df}{f}$$
$$\sin\mu = 0.993,$$
$$\Delta Q = 36\frac{\Delta f}{f} = 0.0036 \text{ for } 1\% \text{ error.}$$

4.1 $$\varepsilon = 2\pi\,\mathrm{mm\,mrad} = 2\pi \times 10^{-6}\,\mathrm{rad}.$$

4.2 $$\frac{a_{\rm h}}{2} = \sqrt{\beta\varepsilon} = 14.7\,\text{mm}$$

$$\gamma = \frac{1+\alpha^2}{\beta} = \frac{1}{\beta},$$

$$x'_{\rm max} = \sqrt{\varepsilon\gamma} = \sqrt{\frac{\varepsilon}{\beta}}$$

$$= 0.136\,\text{mrad}.$$

4.3 $$p = 10\,\text{GeV/c}$$

$$\beta\gamma = \frac{p}{E}\cdot\frac{E}{E_0} = \frac{10}{0.93826} = 10.66$$

$$\varepsilon^* = \varepsilon(\beta\gamma) = 2\times 10.66 = 21.32\pi\,\text{mm mrad}.$$

4.4 $$\beta\gamma = \frac{400}{0.93826} = 426.3,$$

$$\varepsilon = \frac{21.32}{426.3}\pi\,\text{mm mrad} = 50\times 10^{-9}\,\text{m},$$

$$a_{\rm h} = \sqrt{50\times 10^9\times 109} = 2.33\,\text{mm}.$$

4.5 $$p = 10\,\text{GeV/c},$$

$$E = \sqrt{p^2c^2 + m_0^2c^4} = \sqrt{10^2 + 0.938^2} = 10.044\,\text{GeV},$$

$$\frac{v}{c} = \frac{pc}{E} = 0.99562,$$

$$\tau_{\rm rev} = \frac{2\pi R}{\beta c} = \frac{2\pi\times 1100}{0.99562\times 2.9979\times 10^8} = 23.16\,\mu\text{s}$$

$$f_{\rm rev} = 43.186\,\text{kHz}.$$

4.6 Betatron frequencies are $(n \pm Q)f_0$, Q is 6.47; therefore, neighbours to f_0 are $(13-6.47)f_0$, $(1+6.47)f_0$. Difference is $(7.47-6.53)f_0 = 0.06f_0 = 2591$ Hz.

4.7 The vertical bars are spaced in time by the revolution period. The envelope which is $\sin 2\pi f(n-Q)$ takes six revolutions to complete; hence, $n-Q$, the fractional part of Q, is $1/6 = 0.166$.

5.1 $$E^2 = (m_0c^2)^2 + (pc)^2,$$

$$\frac{E^2}{(m_0c^2)^2} = 1 + \frac{(pc)^2}{(m_0c^2)^2} = 1 + \frac{(pc)^2}{E^2}\cdot\frac{E^2}{(m_0c^2)^2},$$

$$\gamma^2 = 1 + \beta^2\gamma^2, \qquad \gamma^2(1-\beta^2) = 1, \qquad \gamma = \sqrt{\frac{1}{1-\beta^2}}.$$

5.2 $$\beta = \sqrt{1 - \frac{1}{\gamma^2}} = \sqrt{1 - \frac{E_0^2}{E^2}}.$$

5.3 $$pc = \sqrt{E^2 - m_0c^2}$$

$$= \sqrt{\gamma^2 - 1}(m_0c^2)$$

$$= (\beta\gamma)m_0c^2.$$

5.4 Differentiating

$$E^2 = E_0^2 + p^2c^2 \;\rightarrow\; 2E\,dE = 2c^2p\,dp,$$
$$dE = v\,dp = \beta c\,dp.$$

Moreover,

$$\frac{pc}{E} = \frac{m_0vc/\sqrt{1-\beta^2}}{m_0c^2/\sqrt{1-\beta^2}} = \beta.$$

Hence,

$$\frac{dE}{E} = \frac{\beta c\,dp}{E} = \frac{\beta c\,dp}{pc/\beta} = \beta^2\frac{dp}{p}.$$

The frequency is a function of velocity and orbit length:

$$f = \frac{v}{2\pi R} \;\Rightarrow\; \frac{df}{f} = \frac{d\beta}{\beta} - \frac{dR}{R} = \frac{d\beta}{\beta} - \alpha\frac{dp}{p}.$$

Find $d\beta/\beta$ by differentiating

$$E = \frac{pc}{\beta} \;\Rightarrow\; \frac{d\beta}{\beta} = \frac{dp}{p} - \frac{dE}{E} = \frac{dp}{p} - \beta^2\frac{dp}{p} = (1-\beta^2)\frac{dp}{p} = \frac{1}{\gamma^2}\frac{dp}{p}.$$

Therefore,

$$\frac{df}{f} = \left(\frac{1}{\gamma^2} - \alpha\right)\frac{dp}{p} = \eta\frac{dp}{p}.$$

5.5 $T = 10$ GeV, $\quad E = 10.9383,$

$E^2 = (0.9383)^2 + (pc)^2,$

$pc = 10.898$ GeV/c,

$B\rho = 3.3356p$ (GeV/c) $= 36.35$ T · m,

$$\rho = \frac{B\rho}{1.5} = 24.23,$$

$R = \frac{3}{2} \times 24.23 = 36.35$ m,

$$\beta = \frac{pc}{E} = \frac{10.898}{10.9383} = 0.9963,$$

$v = \beta c = 0.9963 \times 2.9979 \times 10^8 = 2.9869 \times 10^8,$

$$f_{\text{rev}} = \frac{v}{2\pi R} = 1.308\,\text{MHz}.$$

At 1 GeV

$pc = \{(0.9383 + 1)^2 - (0.9383)^2\}^{1/2} = 1.696,$

$$\beta = \frac{1.696}{1.9383} = 0.8750,$$

$$f_{\text{rev}} = \frac{0.8750 \times 2.9979 \times 10^8}{2\pi \times 36\text{–}35} = 1.148\,\text{MHz},$$

$$fV = 9\,\text{GeV}, \qquad V = \frac{9 \times 10^3}{1.148} = 7.836\,\text{kV},$$

$V \sin 45° = 7.836,$

therefore, $V = 11.04\,\text{kV}$.

5.6 (a) $\dfrac{1}{\gamma_{\text{tr}}^2} = \oint \dfrac{D(s)\,ds}{\rho(s)} \approx \dfrac{\bar{D}}{\bar{\rho}} = \dfrac{9}{36.35}, \quad \gamma_{\text{tr}} = 2.$

(b) $E = 2 \times 0.93827 = 1.886, \qquad pc = \sqrt{1.866^2 - 0.93827^2} = 1.63593\,\text{GeV}.$
(Just below injection!)

(c) $T = 1\,\text{GeV}$:

$$\gamma = \frac{1.93827}{0.93827} = 2.06579,$$

$$\eta = \frac{1}{\gamma^2} - \frac{1}{\gamma_{\text{tr}}^2} = \frac{1}{2.06579^2} - \frac{9}{36.35} = -0.01330.$$

$T = 10\,\text{GeV}$:

$$\gamma = \frac{10.93827}{0.93827} = 11.65789,$$

$$\eta = \frac{1}{\gamma^2} - \frac{1}{\gamma_{\text{tr}}^2} = \frac{1}{11.65789^2} - \frac{9}{36.35} = -0.24027.$$

5.7
$$\beta = \sqrt{1 - \frac{1}{\gamma^2}} = \sqrt{1 - \frac{1}{2.06579^2}} = 0.87503,$$

$$f_s = \sqrt{\frac{|\eta| e V_0}{2\pi E_0 \beta^2 \gamma h}} f_{\text{rf}}$$

$$= \sqrt{\frac{0.01330 \times 7.836 \times 10^3}{2\pi \times 0.93827 \times 10^9 \times 0.76567^2 \times 2.06579 \times 10}} 11.48\,\text{MHz}$$

$$= 438\,\text{Hz}.$$

5.8 See figure below.

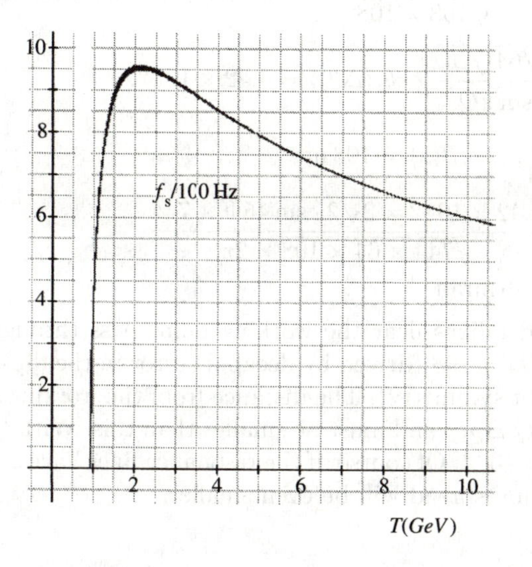

6.1 $\sqrt{\beta\varepsilon} = 28\,\text{mm}, \qquad \hat{\beta} = 104\,\text{m},$

$$\varepsilon = 28^2/104\pi\,\text{mm mrad} = 7.53\,\text{mm mrad},$$

$$\frac{h}{2} = \sqrt{\beta\varepsilon} = \sqrt{20 \times 7.53 \times 10^{-6}} = 12.27 \times 10^{-3},$$

$$= 12.27\,\text{mm},$$

$$x' = \sqrt{\frac{\varepsilon}{\beta}} = \sqrt{\frac{7.53 \times 10^{-6}}{104}} = 0.269\,\text{mrad},$$

$$\frac{\Delta(Bl)}{B\rho} = 2x' = \frac{20 \times B}{1334} = 0.538\,\text{mrad},$$

$$B = 0.035\,\text{T}.$$

6.2 See Fig. 6.7.

6.3 (a) $\dfrac{\Delta(Bl)}{B\rho} = kl\Delta y = 3.22 \times 0.015 \times 0.15 \times 10^{-3} = 7.25 \times 10^{-6}\,\text{rad},$

$$a = \frac{\beta}{2\sin\pi Q} \cdot 7.25 \times 10^{-6}$$

$$= \frac{108 \times 7.25 \times 10^{-6}}{2\sin\pi \cdot 27.75} = 0.55\,\text{mm}.$$

(b) Distortion $= \sqrt{108} \cdot 0.55\,\text{mm} \times \dfrac{1}{\sqrt{2}} = 4.06\,\text{mm} \times 2$ for safety

(average of $\cos\theta$)

$$= 8.12\,\text{mm}.$$

(c) $y(s) = \dfrac{\sqrt{\beta(s)\bar{\beta}}}{2\sqrt{2}\sin\pi Q} \cdot \sqrt{N} \cdot \dfrac{\Delta(Bl)}{B\rho}.$

The increase is by a factor of

$$\frac{\sqrt{108.108 + 108.20}}{\sqrt{108 \times 108}} = 1.089, \quad \text{that is, } 4.42\,\text{mm} \times 2.$$

(d) $\dfrac{\sqrt{744} \cdot \sqrt{64}\sqrt{\beta(s)}}{2\sqrt{2}\sin\pi Q} \cdot \theta_i \langle\Delta\theta_\text{e}\rangle = 4.42 \times 10^{-3},$

$$\theta_i = \frac{2\pi}{744}, \qquad \beta(s) = 108\,\text{m},$$

$$\langle\theta_\text{e}\rangle = \frac{4.42 \times 10^{-3} \times 2\sqrt{2} \cdot \sin 45\Phi \times 744}{\sqrt{744 \times 64 \times 108} \times 2\pi}$$

$$= 0.46\,\text{mrad}.$$

6.4 The fifth-order lines near the working point pass through 27.6, 27.6. The nearest is $5Q_\text{h} = 138$ (driven by decapole error in the dipole magnets which have the right symmetry). The distance from the working point is 0.01. The Q spread is $Q'\Delta p/p$ and must be smaller than this. With $Q' = 36$ this limits $\Delta p/p$ to 2.7×10^{-4}. Of course, if the resonance lines have a comparable width the momentum spread will be even smaller.

6.5
$$Q' = \frac{1}{4\pi}\int \frac{B''(s)\beta(s)D(s)\,ds}{B\rho}$$
$$= \frac{1}{4\pi}\cdot\frac{6\times B''\times 109\times 1.2}{1334} = 10$$
$$B'' = 213.5\,\mathrm{T/m^2}.$$

6.6
$$B = \tfrac{1}{2}B''x^2,$$
$$B = \tfrac{1}{2}B''a^2 = \tfrac{1}{2}213\times(0.08)^2 = 0.681$$

We can increase this by a factor of 1.17 to handle a $B\rho$ of 1565 (momentum 470 GeV/c).

6.7 The calculation of this effect is as before except that D is smaller and, hence, B'' must be larger by a factor of 6. We can achieve this if the aperture is 2×0.033 m.

7.1
$$B\rho = 3.356\times p\ (\mathrm{GeV/c}) = 1334,$$
$$k_{\mathrm{f}} = \frac{1}{B\rho}\frac{\partial B_y}{\partial x} = \frac{19.4}{1334} = 0.01454,$$
$$k_{\mathrm{d}} = -0.01454,$$
$$\Delta Q_{\mathrm{h}} = \frac{1}{4\pi}\int \beta(s)k(s)\frac{\Delta k(s)}{k}\,ds = \sum \frac{\beta_i k_i}{4\pi}\left(\frac{\Delta k}{k}\right)_i l_i$$
$$= \frac{1}{4\pi}\left[108\times0.01454\times3.22\left(\frac{\Delta k}{k}\right)_{\mathrm{HF}} - 18\times0.01454\times3.22\left(\frac{\Delta k}{k}\right)_{\mathrm{HD}}\right]$$
$$= 0.4024\left(\frac{\Delta k}{k}\right)_{\mathrm{HF}} - 0.0671\left(\frac{\Delta k}{k}\right)_{\mathrm{HD}},$$
$$\Delta Q_{\mathrm{v}} = -0.0671\left(\frac{\Delta k}{k}\right)_{\mathrm{HF}} + 0.4024\left(\frac{\Delta k}{k}\right)_{\mathrm{HD}},$$
$$\begin{pmatrix}\Delta Q_{\mathrm{h}}\\ \Delta Q_{\mathrm{v}}\end{pmatrix} = \begin{pmatrix}0.4024 & -0.0671\\ -0.0671 & +0.4024\end{pmatrix}\begin{pmatrix}(\Delta I/I)_{\mathrm{HF}}\\ (\Delta I/I)_{\mathrm{HD}}\end{pmatrix}.$$

Inverting,
$$\begin{pmatrix}(\Delta I/I)_{\mathrm{F}}\\ (\Delta I/I)_{\mathrm{D}}\end{pmatrix} = \begin{pmatrix}2.556 & 0.426\\ 0.426 & 2.556\end{pmatrix}\begin{pmatrix}\Delta Q_{\mathrm{h}}\\ \Delta Q_{\mathrm{v}}\end{pmatrix}.$$

7.2
$$\Delta Q_{\mathrm{h}} = \frac{1}{4\pi}\int \beta(s)\Delta k(s)\,ds,$$
$$\Delta k = -k\left(\frac{\Delta p}{p}\right)$$
$$= \frac{1}{4\pi}\times 108\times 108\times 0.014\times 3.2\left[-\frac{\Delta p}{p}\right],$$
$$Q' = -41.58.$$

7.3
$$\frac{\partial B_y}{\partial x}\ \text{is the quadrupole gradient} = B''x,$$
$$\Delta k = \frac{B''x}{B\rho}.$$

7.4
$$\Delta Q_{\mathrm{h}} = \int \frac{\beta(s)\Delta k(s)}{4\pi}\, ds = \frac{36 \times 108 \times B''x \times 0.85}{4\pi \times 1334}.$$
$$B'' = 150, \qquad x = 2.2 \times 10^{-3}, \qquad \Delta Q_{\mathrm{h}} = 6.5 \times 10^{-2}.$$

7.5
$$\text{Chromaticity} = \frac{dQ_{\mathrm{h}}}{dp/p} = \frac{6.5 \times 10^{-2}}{10^{-3}} = 65 \text{ units}.$$

7.6 Vertical chromaticity (remember $\Delta k = -B''x/B\rho$),
$$\Delta Q_{\mathrm{v}} = -\frac{36 \times 18 \times B''x \times 0.85}{4\pi \times 1334} = 1.084 \times 10^{-2},$$
$$\frac{dQ_{\mathrm{v}}}{dp/p} = -10.8 \text{ units}.$$

7.7 Another set near the D quads will have a predominant effect on vertical chromaticity by virtue of the larger β_{v}.

7.8 4.

7.9
$$\phi = \phi_0 r^4 \sin 4\theta,$$
$$B_x = \frac{\partial\phi}{\partial x} = \cos\theta \cdot \phi_0 4r^3 \sin 4\theta - \sin\theta r^3 4\cos 4\theta$$
$$= 4\phi_0 r^3 \sin 3\theta,$$
$$B_y = \frac{\partial\phi}{\partial y} = \sin\theta \cdot \phi_0 4r^3 \sin 4\theta + \cos\theta r^3 4\cos 4\theta$$
$$= 4\phi_0 r^3 \cos 3\theta,$$
$$\sin 3\theta = \sin(2\theta + \theta) = \cos 2\theta \sin\theta + \cos\theta \sin 2\theta$$
$$= 3\cos^2\theta \sin\theta - \sin^3\theta,$$
$$x = r\cos\theta, \qquad y = r\sin\theta y,$$
$$B_x = 4\phi_0(3x^2y - y^3),$$
$$\cos 3\theta = \cos(2\theta + \theta) = 2\theta\cos\theta - \sin 2\theta \sin\theta$$
$$= \cos^3\theta - \cos\theta\sin^2\theta - 2\sin^2\theta\cos\theta$$
$$= \cos^3\theta - 3\sin^2\theta\cos\theta,$$
$$B_y = 4\phi_0(x^3 - 3xy^2),$$
$$\sin 4\theta = \sin\overline{2\theta + 2\theta} = 2\sin 2\theta\cos 2\theta$$
$$= 2.2\sin\theta\cos\theta 2(\sin^2\theta - \cos^2\theta)$$
$$= 8(\sin^3\theta - \sin\theta\cos^3\theta),$$
$$\phi = 8\phi_0(xy^3 - x^3y).$$

7.10
$$B_y(x) = 4\phi_0(x^3 - 3xy^2)$$
$$\text{when } y = 0, \quad B_y(x) = 4\phi_0 x^3,$$
$$\Delta p = \Delta\beta x',$$
$$\Delta p = \frac{4\phi_0 l\beta}{Bl} a^3 \cos^3\theta,$$

$$2\pi \Delta Q = \frac{\Delta p}{a} \cos\theta,$$

$$\Delta Q = \frac{4\phi_0 \beta l}{2\pi B\rho} a^2 \cos^4\theta$$

$$= \frac{\phi_0 \beta l a^2}{4\pi B\rho} (\cos 4\theta + 4\cos 2\theta + 3),$$

$$\text{width} = \frac{\phi_0 \beta l a^2}{4\pi B\rho}, \qquad \text{shift} = \frac{3\phi_0 \beta l a^2}{4\pi B\rho}.$$

The stopband width depends on amplitude for an octupole.

8.1

$$U_0 = \frac{4}{3} \frac{\pi r_e}{(m_0c^2)^3} \frac{E^4}{\rho}$$

$$= \frac{4}{3}\pi \frac{2.8179 \times 10^{-15}}{(511 \times 10^3)^3} \cdot \frac{(50 \times 10^9)^4}{(27 \times 10^3/2\pi)}$$

$$= 128\,\text{MeV per turn.}$$

At 100 GeV the energy loss is 2^4 times larger, that is, 2.05 GeV/turn.

Note that these answers differ somewhat from Table 8.1 because we assume $\rho = R$.

8.2 The power is $U_0 I = 2.05\,\text{MW}$.

8.3

$$u_c = \left(\frac{h}{2\pi}\right)\omega_c = \frac{3}{4\pi} \frac{hc\gamma^3}{\rho}$$

$$= \frac{3}{4\pi} \cdot \frac{6.6262 \times 10^{-34}}{27 \times 10^3/2\pi} \cdot 2.9979 \times 10^8 \left(\frac{100 \times 10^9}{511 \times 10^3}\right)^3$$

$$= 8.27 \times 10^{-14}\,\text{J},$$

$$\omega = \frac{2\pi u_c}{h} = \frac{8.27 \times 10^{-14} \cdot 2\pi}{6.6262 \times 10^{-34}}$$

$$= 7.84 \times 10^{20},$$

$$\lambda = \frac{2\pi c}{\omega} = \frac{2\pi \times 2.9979 \times 10^8}{7.84 \times 10^{20}}$$

$$= 2.4 \times 10^{-12}\,\text{m}.$$

8.4

$$\frac{\sigma_\varepsilon}{E} = \left[\frac{1.92 \times 10^{-13}}{(27 \times 10^3/2\pi)} \left(\frac{50 \times 10^9}{511 \times 10^3}\right)^2\right]^{1/2} = 0.65 \times 10^{-3}$$

at 100 GeV this is twice as large.

8.5 The mass of the muon is 105.66 MeV/c^2. The classical radius of a particle is inversely proportional to its mass. Hence,

$$r_u = \frac{e^2}{4\pi\varepsilon_0 m_0 c^2} = \frac{0.511}{105.7} \times r_e = 1.36 \times 10^{-17},$$

$$u_0 = \frac{4}{3}\pi \frac{1.36 \times 10^{-17}}{(105.7 \times 10^6)^3} \cdot \frac{(4 \times 10^{12})^4}{4 \times 10^3} = 3.2\,\text{MV}.$$

8.6 $$u_c = \frac{3}{2}\frac{h}{2\pi}\frac{c\gamma^3}{\rho}$$
$$= \frac{3}{4\pi}\cdot\frac{6.6262\times10^{-34}}{4\times10^3}\times 2.9979\times10^8\left(\frac{4\times10^{12}}{106\times10^6}\right)^3$$
$$= 6.37\times10^{-16} = 4\,\text{keV},\ \omega_c = 5.7\times10^{18},\ \lambda = 3.3\times10^{10}.$$

8.7 The damping time $= \dfrac{P_\gamma}{E_0}$
$$= \frac{U_0}{E_0}\cdot\frac{c}{2\pi R}$$
$$= \frac{3.2\times10^6\times2.997\times10^8}{4\times10^{12}\times2\pi\times4\times10^3}$$
$$= 9\times10^{-3}\,\text{s}.$$

9.1 See answer to Exercise 5.5
$$\beta = 0.9963,\qquad \gamma = \frac{10.9383}{0.9383} = 11.657,$$
$$\varepsilon = 10/11.657\pi\,\text{mm mrad} = 8.57\times10^{-1},$$
$$S = \left[\sqrt{8.57\times10^{-7}\times60}\right]^2\pi = 1.62\times10^{-4}m^2,$$
$$\delta Q = \frac{-1.534\times10^{-18}\times1100\times10^{13}}{2\times27.4\times(0.9963)^2(11.657)^3\times1.62\times10^{-4}}$$
$$= 1.2\times10^{-3}.$$

9.2 $$\Omega_0 = 2\pi\times500\times10^6 = 3.14\times10^9,$$
$$\eta = \frac{1}{\gamma^2} - \frac{1}{\gamma_{\text{Tr}}^2} = \frac{1}{(11.657)^2} - \frac{1}{24^2} = 5.62\times10^{-3},$$
$$I_0 = \frac{e\beta c}{2\pi R}\times10^{13} = 0.069\,\text{A},$$
$$E = 10.9383\times10^9\,\text{eV},$$
$$(\Delta\Omega)^2 = -i\left[\frac{5.62\times10^{-3}\times0.069\times(3.14\times10^9)^2}{2\pi(0.9963)^2 10.9383\times10^9}\right]50,$$
$$\Delta\Omega = -i\times1674\,\text{Hz}.$$

9.3 Growth rate $= -\dfrac{2\pi}{\text{Im}(\Delta\Omega)} = \dfrac{2\pi}{1.674} = 3.8\,\text{ms}.$

9.4 $$\xi = \frac{\Delta\Omega^2}{Z/n} = \frac{1.674^2\times10^6}{60}$$
$$= 47\times10^3\,\text{rad}^2/\text{ohm},$$
$$\beta^2 = \left(\frac{1}{\tau}\right)^2 = 69252\,\text{rad}^2,$$
$$Y = \frac{\xi X^2}{4\beta^2} - \frac{\beta^2}{\xi} = \left(\frac{47\times10^3 X^2}{4\times69252} - \frac{69252}{47\times10^3}\right)$$

See figure below.

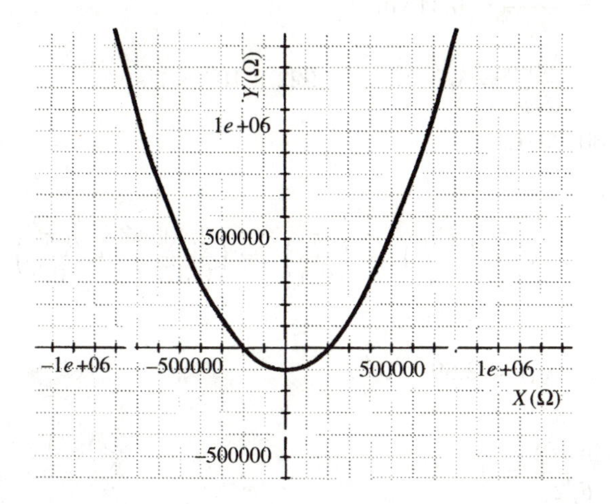

9.5 See figure below.

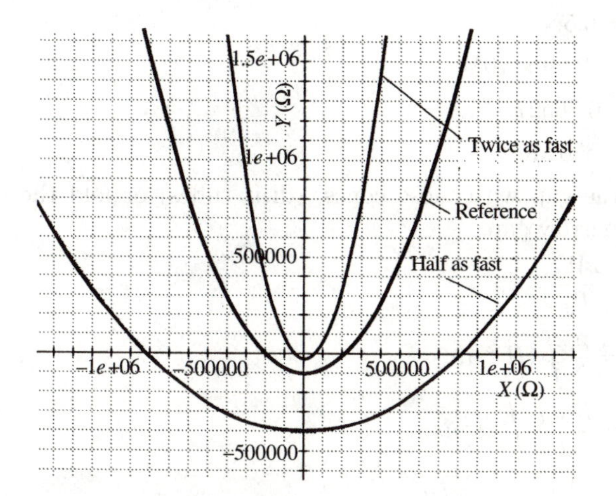

9.6 The diagram is reflected about the real axis.

9.7
$$\left(\frac{\Delta p}{p}\right)^2 = \frac{I_0}{F m_0 c^2 \beta^2 \gamma \eta}\left(\frac{Z}{n}\right)$$

$$= \frac{0.069}{0.938 \times 10^9 \times (0.9963)^2 \times 11.657 \times 5.62 \times 10^{-3}} 50$$

$$\frac{\Delta p}{p} = 2.4 \times 10^{-4}.$$

10.1

$$r_0 = 0.375\,\mathrm{m},$$
$$\Lambda_{010} = \frac{2.405}{r_0} = 6.413\,\mathrm{m}^{-1},$$
$$w_{010} = \frac{\Lambda_{010}}{\sqrt{\varepsilon\mu}} = \Lambda_{010} \times c = 1.922 \times 10^9\,\mathrm{rad/s},$$
$$f = 305\,\mathrm{MHz}.$$

10.2 For TM_{011}

$$\Lambda^2 = \left(\frac{P_{01}}{r_0}\right)^2 + \left(\frac{\pi}{0.5}\right)^2 = \left(\frac{2.405}{0.375}\right)^2 + \left(\frac{\pi}{0.5}\right)^2,$$
$$\Lambda = 8.978\,\mathrm{m}^{-1},$$
$$\omega_{011} = 2.69 \times 10^9.$$

10.3

$$\Gamma = \frac{\sin\theta/2}{\theta/2} = 0.414$$
$$\theta = \frac{4}{3}\pi = \frac{\omega G}{\beta c},$$
$$\omega = 1.922 \times 10^9,$$
$$G = 0.5,$$
$$\beta c = 2.29 \times 10^8$$
$$\beta = 0.76527,$$
$$T = 520\,\mathrm{MeV}.$$

10.4 Reduce the gap with 'nose cones' (drift tubes) or fold the cavity volume about a narrow gap.

10.5

$$Q = \frac{\omega W_\mathrm{s}}{P_\mathrm{d}},$$
$$W_\mathrm{s} = \frac{\mu}{2}\int |H|^2\,dv,$$
$$P_\mathrm{d} = \frac{R_\mathrm{surf}}{2}\int H^2\,ds,$$
$$R_\mathrm{surf} = \frac{1}{\delta\sigma} = \sqrt{\frac{\pi f \mu}{\sigma}}.$$

Therefore, $P_\mathrm{d} = \dfrac{\sqrt{\pi\mu f}}{2\sigma}\displaystyle\int H^2\,ds,$

$$Q = \frac{4\pi f \cdot (\mu/2)\sigma^{1/2} \int H^2\,dv}{\sqrt{\pi\mu f}\int H^2\,ds} = 2\sqrt{\pi f \mu\sigma}\frac{\int |H^2|\,dv}{\int |H^2|\,ds}$$
$$= \frac{2}{\delta}\frac{\int |H^2|\,dv}{\int |H^2|\,ds} = \frac{2}{\delta}K\frac{V}{S}$$

10.6 The points lie on the curve at $\pi/3d$ and $2\pi/3d$. The first point has $v_{\mathrm{ph}} > 0$ while the second is about $0.8c$. The group velocity is less than c but finite for both points. See figure below.

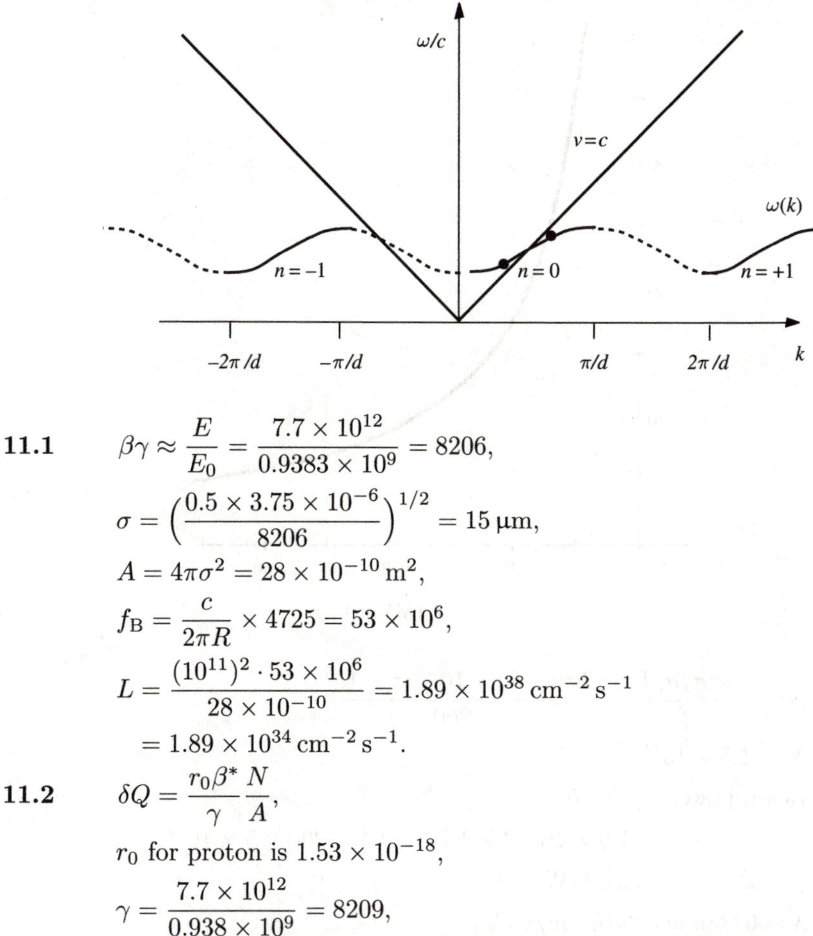

11.1
$$\beta\gamma \approx \frac{E}{E_0} = \frac{7.7\times 10^{12}}{0.9383\times 10^{9}} = 8206,$$
$$\sigma = \left(\frac{0.5\times 3.75\times 10^{-6}}{8206}\right)^{1/2} = 15\,\mu\mathrm{m},$$
$$A = 4\pi\sigma^2 = 28\times 10^{-10}\,\mathrm{m}^2,$$
$$f_{\mathrm{B}} = \frac{c}{2\pi R}\times 4725 = 53\times 10^{6},$$
$$L = \frac{(10^{11})^2\cdot 53\times 10^{6}}{28\times 10^{-10}} = 1.89\times 10^{38}\,\mathrm{cm}^{-2}\,\mathrm{s}^{-1}$$
$$= 1.89\times 10^{34}\,\mathrm{cm}^{-2}\,\mathrm{s}^{-1}.$$

11.2
$$\delta Q = \frac{r_0\beta^*}{\gamma}\frac{N}{A},$$
r_0 for proton is 1.53×10^{-18},
$$\gamma = \frac{7.7\times 10^{12}}{0.938\times 10^{9}} = 8209,$$
$$\delta Q = \frac{1.53\times 10^{-18}\times 0.5}{8209}\,\frac{10^{11}}{28\times 10^{-10}} = 0.0033.$$

11.3 We must reduce N by a factor of 4 and increase the number of buckets which we fill by a factor of 4. Inevitably the luminosity is then reduced by a factor of 4.

11.4 The curve should be the rectangular hyperbola,
$$\frac{l}{2} = \frac{1}{0.7}\,\frac{500(\times 10^{9}\,\mathrm{eV})}{E(\mathrm{MV/m})\times 1000} = 12.5\,\mathrm{km},$$
$$E = 57\,\mathrm{MV/m},$$

$$\lambda = 10\,\text{cm},$$
$$f = 3\,\text{GHz}.$$

See figure below.

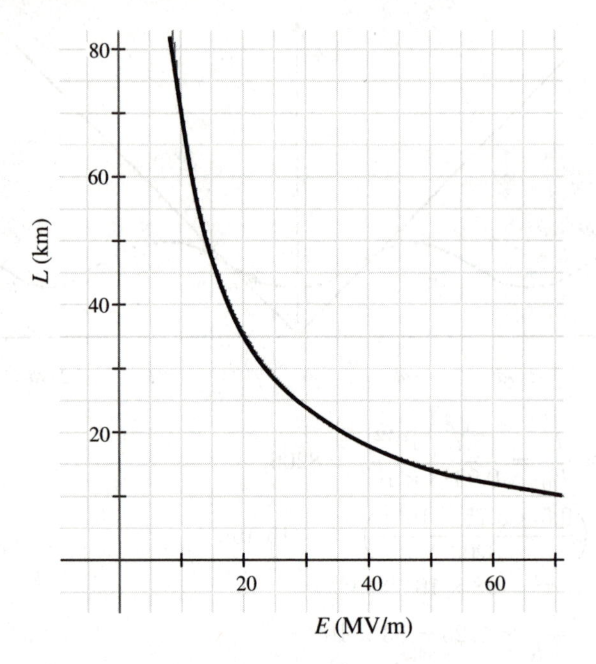

11.5 $$N^2 = \frac{4\pi\sigma_x\sigma_y L}{f} = \frac{4\pi \times (60 \times 10^{-9})^2 \times 10^{38}}{200}$$

$N = 1.5 \times 10^{11}$.

11.6 Beam power $= eNfE$

$$= 1.6 \times 10^{-19} \times 1.5 \times 10^{11} \times 200 \times 5 \times 10^{11}$$
$$= 2.4\,\text{MW}$$

11.7 $\lambda = 6233p \approx 6233E$ (m/GeV)

$= 3.133 \times 10^5$ m at 50 GeV,

$$\tau = \frac{\lambda}{c} = 1.05\,\text{ms}$$

$= 2.5 \times 10^7$ m at 4 TeV = 83 msec.

$B\rho = 1.33 \times 10^4$ T m,

assume $\rho/R = 0.7$

$$R = \frac{1.33 \times 10^4}{0.7 \times 6} = 3.26 \text{ km},$$

$2\pi R = 20$ km,

$Q \approx 70$.

Number of periods = 280 for $\mu = 90°$.

Period length = 71 m. Say 8 bending magnets of 6.5 m plus 2 quadrupole.

11.8 Decay time is 83 ms = 1/12 second. Repetition rates of 10–15 Hz are appropriate.

12.1
$$\frac{1}{\tau} = \frac{2W}{N}[2g - g^2(1+\rho)],$$
$$\frac{d(1/\tau)}{dg} = \frac{2W}{N}[2 - 2g(1+\rho)],$$
$$g = \frac{1}{1+\rho}, \quad \text{that is, somewhat less than 1,}$$
$$\frac{1}{\tau} = \frac{2W}{N(1+\rho)}.$$

12.2
$$\text{Sample time} = \frac{1}{300 \times 10^6} = 3\,\text{ns}.$$
Revolution time is $2\pi R/v = 546\,\text{ns}$.
$$\text{Sample is } \frac{3\,\text{ns}}{546\,\text{ns}} \times 10^6 = 5497 \text{ particles}.$$

12.3
$$m_e c^2 = 0.511 \times 10^6\,\text{eV}.$$
$$T = 60 \times 10^3\,\text{eV},$$
$$E_e = 571\,\text{MeV},$$
$$\gamma_e = \frac{571}{511} = 1.117,$$
$$\beta_e = \sqrt{1 - (1/\gamma^2)} = 0.446.$$
Proton with same β and γ,
$$E_p = 1.117 \times m_p c^2 = 1.05\,\text{GeV}.$$
$$T = 110\,\text{MeV}, \qquad p = 467\,\text{MeV/c}.$$

12.4
$$x' = \frac{40 \times 10^{-6}}{5 \times 10^{-2}} = 0.8\,\text{mrad}.$$
The electron beam alignment must be better than this.

12.5 There are various ways to simulate cooling but with a small number of particles the result often oscillates.

12.6
$$x' = \sqrt{\frac{2 \times 10^{-6}}{10}} = 0.45 \times 10^{-3},$$
Transverse velocity is $c \times 0.45 \times 10^{-3} = 1.35 \times 10^5\,\text{m/s}$.

12.7
$$\tfrac{1}{2}mv^2 = \tfrac{3}{2}kT,$$
$$\tfrac{1}{2} \times 1.672 \times 10^{-27} \times (1.35 \times 10^5)^2 = \tfrac{3}{2} \times 1.38 \times 10^{-23} T,$$
$$T = 726\,000\,\text{K}.$$

12.8 $\lambda = 5485 \times 10^{-10}\,\text{m}$,

$$\Delta E = \hbar\omega = 6.58 \times 10^{-16}\,\text{eV} \times \frac{2\pi c}{5.485 \times 10^{-7}}$$
$$= 2.26\,\text{eV}.$$

13.1 $\text{Beam intensity} = \dfrac{5\,\text{mA}}{1.6 \times 10^{-19}} = 3.13 \times 10^{16}\,\text{s}^{-1}$,

Beam power $= 5\,\text{mA} \times 2\,\text{MV} = 10\,\text{kW}$.

13.2 $\text{Area swept} = \dfrac{15\,\text{cm} \times 80\,\text{cm}}{60\,\text{s}} = 20\,\text{cm}^2\,\text{s}^{-1}$,

Particles $\text{s}^{-1} = 3.13 \times 10^{16}$,

Energy deposited per particle.

$2.1\,\text{MeV}/(\text{g}\,\text{cm}^{-2}) \times 0.20 \times 1.4 = 588\,\text{keV} = 9.4 \times 10^{-14}\,\text{J}$

Power $= 2.95\,\text{kW}$.

13.3 Mass receiving this power $1.4\,\text{g}\,\text{cm}^{-3} \times 20\,\text{cm}^2 \times 0.2\,\text{cm} = 0.56\,\text{kg}$.

Energy received $= 525\,\text{kW}\,\text{s}\,\text{kg}^{-1}$, that is, $525\,\text{kGy}$.

13.4 1 Å is 0.1 nm which is the scale of atoms. Crystals, molecules, and DNA are in the range 1–10 nm.

13.5 $$pc = \hbar\omega = \frac{6.58 \times 10^{-16} \times 2.998 \times 10^{8} \times 2\pi}{10^{-10}} = 12.4\,\text{keV},$$

$$kE = \frac{p^2}{2m} = \frac{(12.4 \times 10^3)^2}{2 \times 0.938 \times 10^9} = 80\,\text{meV}.$$

13.6 $$\frac{25 \times 0.4}{10^9 \times 1.6 \times 10^{-19}} = 62 \times 10^{15}\,\text{neutrons s}^{-1}$$

This is two orders of magntiude stronger than the reactor. However, the proton burst is 0.4 μs at 50 Hz leading to a mean flux of 10^{12} ns^{-1}.

14.1 $$\frac{30 \times 10^3 \times 7 \times 10}{0.5} = 4.2\,\text{MV/m}.$$

14.2 $$C = \frac{A\varepsilon_0}{d} = \frac{\pi \times 0.7^2 \times 8.85 \times 10^{-12}}{0.05} = 2.72 \times 10^{-10}\,\text{F}$$

$V = 30 \times 10^3$

Energy stored $= \frac{1}{2}\,CV^2 = 0.12261\,\text{J}$

Intensity $= N$.

$$\begin{aligned}\text{Change in beam energy} &= Ne\Delta v \\ &= N \times 1.6 \times 10^{-19} \times 210 \times 10^3 \\ &= 0.3 \times 0.12261\end{aligned}$$

$N = 1.1 \times 10^{12}$

14.3 $$\Delta\omega = \omega_1 - \omega_2 = \frac{2\pi c}{\lambda_1} - \frac{2\pi c}{\lambda_2}$$
$$= 2\pi \times 2.9979 \times 10^8 \left(\frac{1}{9.6 \times 10^{-6}} - \frac{1}{10.27 \times 10^{-6}}\right)$$
$$= 1.28 \times 10^{13}\,\mathrm{s}^{-1}.$$

14.4 $$\omega_\mathrm{p} = \left(\frac{n_0 e^2}{\varepsilon_0 m_\mathrm{e}}\right)^{1/2},$$
$$n_0 = \frac{\varepsilon_0 m_\mathrm{e}}{e^2}\omega_\mathrm{p}^2$$
$$= \frac{8.854 \times 10^{-12} \times 9.1 \times 10^{-31}}{(1.6 \times 10^{-19})^2}(1.28 \times 10^{13})^2$$
$$= 5 \times 10^{22}\,\mathrm{m}^{-3} \quad (5 \times 10^{16}\,\mathrm{cm}^{-3}).$$

References

Alvarez, L. W. (1946). The design of a proton linear accelerator. *Physical Review*, **70**, 799.

Amaldi, E. (1981). *The Bruno Touschek Legacy*, CERN 81-19.

Amaldi, U. (1987). *Proceedings of the Workshop on Physics at Future Accelerators*, La Thuile, CERN 87–07, Vol. 1, pp. 323–52.

Autin, B. and Bryant, P. J. (1971). Closed orbit manipulation and correction in the CERN ISR. *Proceedings of the 8th International Conference on High Energy Accelerators*, CERN, Geneva.

Baglin, J. E. E., Kellock, A. J., and Tabacniks, M. H. (1996). PIXE for thin film analysis. *Proceedings of the 14th International Conference on Applications of Accelerators in Research and Industry*, Denton, Texas, pp. 563–6. AIP Press, New York.

Barber, W. C., *et al.* (1966). Operation of the electron–electron storage ring at 550 MeV. *Symposium International sur les Anneaux de Collisions, Electron et Positrons*, Saclay, P.U.F., Paris.

Basrour, S., Bernede, P., Casses, V., Kupka, R., Labeque, A., Liu, Z. W., Megtert, S., and Rouliay, M. (1996). Synchrotron radiation for microstructure fabrication. *Proceedings of the 14th International Conference on the Application of Accelerators in Research and Industry*, Denton, Texas, pp. 745–8. AIP Press, New York.

Bechtold, V. (1996). Isotope production with cyclotrons. *Proceedings of the CERN Accelerator School*, La Hulpe. CERN 96-02, pp. 329–42.

Bethge, K. (1993). *Industrial applications of accelerators in advances of accelerator physics and technologies.* World Scientific, Singapore.

Bethge, K. H. W. (1996a). Applications of ion linacs. *Proceedings of the CERN Accelerator School*, La Hulpe. CERN 96-02, pp. 259–66.

Bethge, K. H. W. (1996b). Non-medical applications of linacs. *Proceedings of the CERN Accelerator School*, La Hulpe. CERN 96-02, pp. 249–58.

Billinge, R. (1984). Design and operation of the CERN SppS collider, Overview, *Proceedings of the 1984 Summer School on High Energy Particle Accelerators*, FNAL.

Blewett, M. H. (1977). Theoretical aspects of the behaviour of beams in accelerators and storage rings. CERN Report 77-13, pp. 111–38.

Bovet, C., Gouiran, R., Gumowski, I., and Reich, H. (1970). A selection of formulae and data useful for the design of A.G. synchrotrons. CERN/MPS-SI/Int.DL/70/4.

Brinkmann, R. (1987). Insertions. *Proceedings of the 1986 CERN Accelerator School*, CERN 87-10.

Bruck, H. (1966). *Accélérateturs circulaires de particules.* Presses Universitaires de France, Paris.

Bryant, P. J. (1992). Insertions. *Proceedings of the CERN Accelerator School, Jyväskylä*, Finland, 1992, CERN 94-01, Vol. 1.

Bryant, P. J. and Johnsen, K. (1993). *Circular accelerators and storage rings.* Cambridge University Press.

Budker, G. I. (1964). *Creation of accelerators with colliding beams.* Vestnik Akademii Nauk SSSR No. 6.

Budker, G. I. (1967). An effective method of damping particle oscillations in proton and antiproton storage rings. *Soviet Journal of Atomic Energy*, **22**, 438.

CERN, Geneva (1984). *LEP Design Report*, CERN-LEP/84-01.

Chao, A. W. (1993). *Physics of collective beam instabilities in high energy accelerators.* Wiley, New York.

Chao, A. W. and Tignier, M. (1998). *Handbook of accelerator physics.* World Scientific, Singapore, p. 340.

Christofilos, N. C. (1950). Unpublished report and U.S. Patent no. 2.736,799, filed March 10, 1950, issued February 28 1956.

Cleland, M. R. (1987). *Proceedings of linear accelerator conference.* IEEE, Piscataway.

Cockcroft, J. D. and Walton, E. T. S. Experiments with high velocity ions. *Proceedings of the Royal Society*, London. **A136**, 619 (1932); **A137**, 229 (1932); **A144**, 704 (1934).

Collin, R. E. (1961). *Field theory of guided waves.* McGraw-Hill, New York.

Collin, R. E. (1966). *Foundations for microwave engineering.* International Student Edition, McGraw-Hill, New York.

Conte, M. and MacKay, W. W. (1991). *An introduction to the physics of particle accelerators.* World Scientific, Singapore.

Courant, E. D. and Snyder, H. S. (1958). Theory of the alternating-gradient synchrotron. *Annals of Physics*, **3**, 1–48.

Courant, E. D., Livingston, M. S., and Snyder, H. S. (1952). The strong-focusing synchrotron—a new high-energy accelerator. *Physical Review*, **88**, 1190–6.

Dawson, J. M. and Chen, P. (1985). SLAC-PUB-3601.

Dearnaley, G. (1987). *Nucl. Inst. & Meth.* **B24/25**, 506.

Degèle, D. (1980). PETRA. *Proceedings of the 8th International Conference on High Energy Accelerators*, Geneva. Birkhaüser, Basel.

Delahaye, J. -P. *et al.* (1999). CERN/PS 99-005 (LP), 1999 Paricle Accelerator Conference, New York, Eds. A. Luccia, W. Mackay. IEEE, Piscataway.

DESY, Hamburg (1981). HERA—A proposal for a large electron–proton colliding beam facility. DESY HERA 81/10.

Drewell, N. H., Lawrence, C. B., and McKeown, J. (1996). The IMPELA accelerator, field performance and new developments. *Proceedings of the 14th International Conference on the Application of Accelerators in Research and Industry*, Denton, Texas, pp. 1077–80. AIP Press, New York.

Edwards, D. and Syphers, M. J. (1993). *An introduction to the physics of high energy accelerators.* Wiley-Interscience, New York.

Elder, F. R., Gurewitsch, A. M., Langmuir, R. V., and Pollock H. C. (1947). A 70 MeV synchrotron. *Journal of Applied Physics*, **18**(9), 810–18.

Evans, L. (1984). Design and operation of the CERN SppS collider. *Proceedings of the 1984 Summer School on High Energy Particle Accelerators,* FNAL.

Feynman, R. P. (1963). *Lectures on Physics.* Addison-Wesley Publ. Co. Inc., Reading, Mass, Vol. II, p. 28.

Gahbauer, R. A. and Wambersie, A. (1996). Medical applications of electron linear accelerators. *Proceedings of the CERN Accelerator School*, La Hulpe, CERN 96-02, pp. 229–48.

Garren, A. A., Kenney, A. S., Courant, E. D., and Syphers, M. J. (1985). SYNCH, *Fermilab Report* FN 420.

Goward, F. K. and Barnes, D. E. (1946). *Nature*, **158**, 413.

Griffin, E. (ed.) (1980). The Fermilab saver/collider and future options. *Proceedings of the 11th International Conference on High Energy Accelerators*, pp. 92–102, Birkhaüser, Basel.

Grob, J. J., Muller, D., and Stoquert, J. P. (1996). Recent advances in MeV ion implantation. *Proceedings of the 14th International Conference on the Application of Accelerators in Research and Industry*, Denton, Texas, AIP Press, New York, pp. 1001–4.

Guiducci, S. (1992). Chromaticity. *Proceedings of the CERN Accelerator School, Jyväskylä*, Finland, 1992, CERN 94-01, Vol. 1.

Guignard, G. (1970). Effets des champs magnetiques perturbateurs d'un synchrotron. CERN Report 70-24, pp. 67–105.

Guignard, G. (1978). A general treatment of resonances in accelerators. CERN Report 78–11.

HERA Design Team (1981). HERA a proposal for a large electron proton colliding beam facility at DESY. DESY HERA 81/10.

Hereward, H. G. (1976). Landau damping. *Theoretical aspects of the behaviour of beams in accelerators and storage rings. International School of Particle Accelerators*, Erice, CERN 77-13.

Hofmann, A. (1976). Single beam collective phenomena—longitudinal. *Theoretical aspects of the behaviour of beams in accelerators and storage rings. International School of Particle Accelerators*, Erice. CERN 77-13.

Hofmann, A. (1992). Tune shifts from self-fields and images. *Proceedings of the CERN Accelerator School, Jyväskylä, Finland 1992*, CERN 94-01, Vol. 1.

Hofmann, I. (1996). Application of RF-linacs to heavy-ion fusion. *Proceedings of the CERN Accelerator School*, La Hulpe. CERN 96-02, pp. 267–84.

Hübner, K. (1984). *Proceedings of the CERN Accelerator School*, Gif-sur-Yvette, 1984, CERN 85-19, Vol. 1.

Huxley, L. G. H. (1943). *A survey of the principles and practice of wave guides.* Cambridge University Press.

Iselin, F. C. and Grote, H. G. (1991). The MAD program, Version 8.4, CERN/SL/90-13.

Ising, G. (1924). Prinzip einer Methode zur Herstellung von Kanalstrahlen höher Voltzahl. *Arkiv för matematik o. fysik*, **18**(30), 1–4.

Jackson, J. D. (1962). *Classical electrodynamics.* Wiley, New York.

Jianjun, W., Jiarong, Y., Kexin, L., Kun, L., Qiang, Z., Shijun, G., Sixun, Y., Xiangyang, L., and Zimo, Z. (1996). Applications of AMS radiocarbon dating in Chinese archaeological studies. *Proceedings of the 14th International Conference on the Application of the Accelerator in Research and Industry,* Denton, Texas, pp. 803–6. AIP Press, New York.

Johnsen, K. (ed.) (1964). The design study of intersecting storage rings (ISR), CERN AR/Int. SG/64-9.

Judd, D., private communication.

Keil, E. and Schnell, W. (1969). CERN Report ISR-TH-RF/69-488.

Kerst, D. W. and Serber, R. (1941). Electronic orbits in the induction accelerator. *Physical Review*, **60**, 53–8.

Kerst, D. W., Cole, F. T., Crane, H. R., Jones, L. W., Laslett, L. J., Ohkawa, T., Sessler, A. M., Symon, K. R., Terwilliger, K. M., and Vogt Nilsen, N. (1956). Attainment of very high energy by means of intersecting beams of particles. *Physical Review*, **102**, 590–1.

Laclare, J. L. (1980). Bunched-beam instabilities. *11th International Conference on High Energy Accelerators*, Geneva. Birkhaücer, Basel. p. 526.

Laclare, J. L. (1982). Instabilities in storage rings. *Proceedings of the Symposium on Accelerator Aspects of HI Fusion*, Darmstadt FRG, GSI Report 82–8.

Laclare, J. L. (1992). Coasting beam transverse coherent instabilities. *Proceedings of the CERN Accelerator School, Jyväskylä*, Finland, 1992. CERN 94-01, Vol. 1.

Lapostolle, P. M. (1986). Proton linear accelerators. *Los Alamos Report* LA-11601-MS.

Lapostolle, P. and Septier, A. (1970). *Linear accelerators.* North-Holland, Amsterdam.

Laslett, L. J. and Resegotti, L. (1967). *Proceedings of the 6th International Conference on High-Energy Accelerators*, Cambridge, USA (Cambridge Electron Accelerator, Cambridge, MA), p. 150.

Lawrence, E. O. and Edelfsen, N. E. (1930). *Science*, **72**, 376–7.

Lawson, J. D. (1970). *Contemporary Physics*, **11**(6), 575.

Le Duff, J. (1992). Longitudinal beam dynamics in circular accelerators. *Proceedings of the CERN Accelerator School, Jyväskylä*, Finland, 1992. CERN 94-01, Vol. 1.

Lefèvre, P. (ed.) (1995). *The Large Hadron Collider*, CERN/AC/95-05.

Lengeler, H. (1996). Application of high intensity linacs to spallation sources. *Proceedings of the CERN Accelerator School*, La Hulpe. CERN 96-02, pp. 285–96.

Lewis, D. M. (1996). Designing a radioisotope facility. *Proceedings of the CERN Accelerator School*, La Hulpe. CERN 96-02, pp. 343–62.

Livingood, J. J. (1961). *Principles of cyclic particle accelerators.* Van Nostrand, New York.

Livingston, M. S. and Blewett, J. P. (1962). *Particle accelerators.* McGraw-Hill, New York.

Loew, G. A. (1983). Elementary principles of linear accelerators. *SLAC PUB* 3221. Also in *AIP Conference Proceedings* No. 105, New York.

Mandrillon, P., Besançon, S., Fiétier, N., Giusto, A., Michel, S., Ponnelle, S., and Rubbia, C. (1996). Important design issues of a high-efficiency cyclotron complex to drive the energy amplifier. *5th European Particle Accelerator Conference*, EPAC 96, Sitges, Barcelona. Institute of Physics, Bristol.

McMillan, E. M. (1945). The synchrotron—a proposed high energy accelerator. *Physical Review*, **68**, 143.

Møller, S. P. (1992). Cooling techniques. *Proceedings of the CERN Accelerator School, Jyväskylä*, Finland, 1992. CERN 94-01, Vol. II, pp. 601–16.

Montague, B. W. (1977). RF acceleration. *Theoretical Aspects of the Behaviour of Beams in Accelerators and Storage Rings: International School of Particle Accelerators*, Erice. CERN 77-13.

Montgomery, C. G., Dicke, R. H., and Purcell, E. M., (1948). *Principles of microwave circuits*, M.I.T. Radiation Lab. Series, McGraw-Hill, New York.

Moreno, T. (1958). *Microwave transmission design data.* Dover, New York.

Mutsaers, P. H. A., Rokita, E., and de Voigt, M. J. A. (1996). Charged particle beam and synchrotron radiation as supplementary tools in biomedical studies. *Proceedings of the 14th International Conference on Applications of Accelerators in Research and Industry*, Denton, Texas, AIP Press, New York, pp. 749–52.

Oliphant, M. (1967). The genesis of the Nuffield cyclotron and the proton synchrotron. Department of Physics, University of Birmingham.

Paterson, J. M. (1980). P.E.P. *Proceedings of 8th International Conference on High Energy Accelerators*, Geneva.

Prior, C. R. (1998). Status of the HIDIF Study. *Proceedings of the 6th European Particle Accelerator Conference*, EPAC 98, Stockholm, Sweden, Institute of Physics, Bristol.

Ramo, S., van Duzer, T., and Whinnery, J. R. (1984). *Fields and waves in communication electronics.* 2nd ed., Wiley, New York.

Rees, G. H. (1989). Radiation excitation and beam distributions in electron storage rings. *Proceedings of the CERN Accelerator School*, Chester. CERN 90-03, Vol. I.

Richter, B., *et al.* (1980). *SLC design report*, SLAC 2590.

Rossbach, J. and Schmüser, P. (1992). Basic course on accelerator optics. *Proceedings of the CERN Accelerator School, Jyväskylä*, Finland, 1992. CERN 94-1, Vol. I. pp. 17–88.

Rubbia, C. (1996). The energy amplifier: a discription for the non-specialists. CERN/ET/International Note 96-01.

Ruth, R. D., Chao, A. W., Morton, P. L., and Wilson, P. B. (1984). SLAC-PUB-3374.

Sacherer, F. (1972). *CERN Report* SI/BR/72-5.

Sarantsev, V. P. and Ivanov, I. N. (1981). *Proceedings of the 4th International Topical Conference on High-Processing Beams*, Palaiseau, p. 691.

Schachinger, L. and Talman, R. (1985). TEAPOT, SSC-52.

Scharf, W. H. and Chomicki, C. A. (1996). *Physica Medica* XII(4), 199.

Schmüser, P. (1987). Basic course on accelerator optics. *Proceedings of the 1986 CERN Accelerator School*, CERN 87-10.

Servranckx, R. and Brown, K. L. (1984). DIMAD, *SLAC Report* 270 UC-288.

Sisteron, J. M. (1996). Proton therapy in 1996. *Proceedings of the 14th International Conference on Applications of Accelerators in Research and Industry*, Denton, Texas, AIP Press, New York, pp. 1261–4.

Slater, J. C. (1969). *Microwave electronics.* Dover, New York.

Steffen, K. (1984). *High energy beam optics.* Wiley, Introduction to weak and strong focusing. *Proceedings of the CERN Accelerator School, Gif-Sur-Yrette.* CERN 85-19, Vol. I.

Symon, K. R., *et al.* (1956). Fixed field alternating-gradient particle accelerators. *Physical Review*, **103**, 1837.

Tajima, T. and Dawson, J. M. (1979). *Phys. Rev. Lett.*, **43**, 267.

Tajima, T. and Dawson, J. M. (1982). *Proceedings of the Workshop on Laser Acceleration of Particles*, Los Alamos, AIP Conf. Proc. No. 91, p. 69.

van de Graaff, R. J. (1931). A 1 500 000 Volt electrostatic generator. *Physical Review*, **38**, 1919–20.

van der Meer, S. (1972). Stochastic damping of betatron oscillations in the ISR, CERN/ISR-PO/72-31.

van der Meer, S. (1984). Stochastic cooling. *Proceedings of the 1984 Summer School on High Energy Particle Accelerators*, FNAL.

van der Meer, S. (1985). CERN-PS-85-65.

Veksler, V. I. (1944). A new method of accelerating relativistic particles. *Comptes Rendus (Doklady) de l'Academie des Sciences de l'URSS*, **43**(8), 329–31.

Voss, G. A. and T. Weiland, T. (1982). Particle acceleration by wake fields, DESY M-82-10.

Walker, R. P. (1992a). Radiation damping. *Proceedings of the CERN Accelerator School, Jyräskylä*, Finland, 1992. CERN 94-01, Vol. 1, pp. 461–80.

Walker, R. P. (1992b). Synchrotron radiation. *Proceedings of the CERN Accelerator School, Jyräskylä*, Finland, 1992. CERN 94-01, Vol. 1, pp. 347–459.

Weiland, T. (1985) *Proceedings of the CAS-ECFA-INFN Workshop on Generation of High Fields*, Frascati. CERN 85-07.

Wideröe, R. (1928). Uber ein neues Prinzip zur Herstellung höher Spannungen. *Arch. f. Electrot.*, **21**, 387–406.

Wideröe, R. (1994). *The infancy of particle accelerators*, DESY 94-039.

Wideröe, R. (Original copy books from 1923 to 1928). ETH Library Zürich, **Hs 903**, 633–8.

Willis, W. (1985). *Proceedings of the Workshop on Laser Acceleration of Particles*, Malibu (AIP Conf. Proc. No. 130), p. 421.

Wilson, E. J. N. (ed.) (1972). The 300 GeV Programme, CERN/1050.

Wilson, E. J. N. (1984). Non-linearities and resonances. *Proceedings of the CERN Accelerator School*, Gif-Sur-Yvette, 1984, CERN 85-19. Vol. I.

Wilson, P. B. (1982). High energy electron linacs: applications to storage ring RF systems and linear colliders. *SLAC PUB* 2884. Also in *AIP Conference Proceedings*, No. 87, New York.

Wilson, R. R. (1946). Radiological use of fast protons. *Radiology*, **47**, 487.

Wilson, R. R. (1971). The NAL proton synchrotron. *Proceedings of the 8th International Conference on High Energy Accelerators*, CERN.

Zotter, B. (1976). CERN Report ISR-GS/76-11.

Index